TIMELINES
OF
SCIENCE
AND TECHNOLOGY
VOLUME 4

THE SCIENTIFIC
REVOLUTION

1625–1774

JOHN O.E. CLARK

GROLIER

an imprint of

SCHOLASTIC

www.scholastic.com/librarypublishing

Published by Grolier
an imprint of Scholastic Library Publishing,
Old Sherman Turnpike
Danbury, Connecticut 06816

© 2006 The Brown Reference Group plc

Set ISBN 978-0-7172-6101-7
Volume 4 ISBN 978-0-7172-6105-5

Library of Congress Cataloging-in-Publication Data

Timelines of science and technology
        p. cm.
        Includes bibliographical references and index
        Contents: v. 1. Origins of science — v. 2. Classical and early medieval science — v. 3. Late medieval and Renaissance science — v. 4. The Scientific Revolution — v. 5. The Industrial Revolution — v. 6. The Age of Steam — v. 7. The Age of Electricity — v. 8. The Atomic Age — v. 9. The Space Age — v. 10. The modern world.
        ISBN 978-0-7172-6101-7 (set : alk paper) — ISBN 978-0-7172-6102-4 (v. 1 : alk paper) — ISBN 978-0-7172-6103-1 (v. 2 : alk paper) — ISBN 978-0-7172-6104-8 (v. 3 : alk paper) — ISBN 978-0-7172-6105-5 (v. 4 : alk paper) — ISBN 978-0-7172-6106-2 (v. 5 : alk paper) — ISBN 978-0-7172-6107-9 (v. 6 : alk paper) — ISBN 978-0-7172-6108-6 (v. 7 : alk paper) — ISBN 978-0-7172-6109-3 (v. 8 : alk paper) — ISBN 978-0-7172-6110-9 (v. 9 : alk paper) — ISBN 978-0-7172-6111-6 (v. 10 : alk paper)
        1. Science-History. 2. Discoveries in science. 3. Technology-History. I. Grolier (Firm)

Q125.T587 2006
509—dc22   2005050387

For information address the publisher:
Grolier, Sherman Turnpike,
Danbury, Connecticut 06816

Printed and bound in Singapore.

FOR THE BROWN REFERENCE GROUP

Consultant: Erin Dolan, Virginia Polytechnic and State
    University, U.S.A.

Project Editor: Graham Bateman
Editor: Virginia Carter
Designer: Steve McCurdy
Research: Geoff Roberts
Production: Alastair Gourlay, Maggie Copeland
Editorial Director: Lindsey Lowe

PICTURE CREDITS
(t = top, b = bottom, c = center, l = left, r = right)

Cover: *Space Shuttle*, NASA Headquarters – Greatest Images of NASA; *19th-Century Steam Train*, Science Photo Library; *Galileo*, AKG-London

6c TopFoto; **6b** Fotomas/TopFoto; **7** Science Musuem; **8** Schoenberg Center; **9t** Archivo Iconografico, S.A./Corbis; **9b** Ann Ronan Picture Library/TopFoto; **10b** Science Museum Library; **11t** Science Museum; **12** Science Museum; **14t** NASA Jet Propulsion Laboratory; **14c** Schoenberg Center; **15c** Clayton J. Price/Corbis; **15b** Marc Garanger/Corbis; **16t** Bettmann/Corbis; **16–17** Science Museum Library; **17t** Bettmann/Corbis; **18t** Bettmann/Corbis; **18b** TopFoto; **19b** Science Museum Library; **20t** Ann Ronan Picture Library/TopFoto; **21t** Roger-Viollet/TopFoto; **21b** TopFoto; **22b** Science Museum Pictorial; **23t** The British Library/TopFoto; **23b** Science Museum Library; **24** ©2004 Ann Ronan Picture Library/TopFoto; **24-25** ©2004 Ann Ronan Picture Library/TopFoto; **26b** Science Museum; **27c** ©2003 Charles Walker/TopFoto; **28t** ©2004 Ann Ronan Picture Library/TopFoto; **29t** ©2004 Ann Ronan Picture Library/TopFoto; **30c** ©2004 Ann Ronan Picture Library/TopFoto; **31t** NASA Headquarters – Greatest Images of NASA; **31c** Science Museum; **31b** Science Museum/TopFoto; **32t** ©2004 Ann Ronan Picture Library/TopFoto; **33t** Science Museum; **33b** TopFoto; **34c** ©Topham Picturepoint/TopFoto; **36t** Science Photo Library; **37t** Science Photo Library; **38t** ©2002 ARPL/Topham/TopFoto; **38b** TopFoto; **39b** Science Museum; **40t** Schoenberg Center; **40b** Schoenberg Center; **41** Philadelphia Museum of Art/Corbis; **42c** Schoenberg Center; **42b** Topham Picturepoint/TopFoto; **43** The British Library/TopFoto; **44t** ©2004 TopFoto/TopFoto: **44b** Science Museum; **45** Science Museum Pictorial

The Brown Reference Group has made every effort to trace copyright holders of the pictures used in this book. Anyone having claims to ownership not identified above is invited to contact The Brown Reference Group.

CONTENTS

How to Use this Set

Each of the volumes in this set covers a distinct period in our history. The periods are displayed on pages 6–7 of Volume 1, together with detailed contents. Within each volume there are three types of articles. TIMELINE ARTICLES list year by year the discoveries or inventions. They are arranged in horizontal bands corresponding to particular disciplines so that you can see at a glance how they relate to other areas of scientific knowledge. Each Timeline band—for example, physics or chemistry—has its own color. Interspersed among the Timeline articles are double-page SPECIAL FEATURES, elaborating a particular topic from a Timeline. These in-depth articles focus on the background to a discovery, give information about the people involved, and explain the ways in which the discoveries or inventions have been put to use. For example, in Volume 8 (pages 28–29) "The Evolution of the Helicopter" develops a 1936 Timeline article from page 26 about helicopters. Boxed features add to the available information, often explaining scientific principles. The KEY PEOPLE pages at the end of each volume give full biographical details about prominent individuals or groups mentioned in the Timelines or the Features, thus weaving together in one story their achievements. Used together, the three article styles enable you to build up a comprehensive picture of the circumstances leading to a particular breakthrough moment.

Fully captioned ILLUSTRATIONS play a major role in this set. They include early prints, contemporary photographs, artwork reconstructions, and explanatory diagrams.

A number of cross-reference devices help you navigate through the volumes. Names of individuals with detailed biographies in the Key People pages are highlighted throughout the text using a bold typeface, for example, **Alfred Nobel**. Some individuals lived and worked in periods covered by more than one volume; and sometimes a person from an earlier or later period in history is mentioned. For this reason, a name in a bold typeface will not necessarily be found in the Key People pages of the particular volume being studied. To find out the location of a biography, consult the Set Index beginning on page 53 of each volume. Other names are printed in a bold italic typeface, for example, *Michael Faraday*. This signifies that the person is the subject of a double-page feature article (in this case The Nature of Light, Volume 4, pages 28–29). Again, the exact page reference can be found in the Index.

At the top of most articles you will find direct cross references to special features that are relevant to a topic contained in the article.

A GLOSSARY of some terminology used is included in each volume. It will help you if there are words that you do not fully understand. Each volume ends with a list of FURTHER READING AND USEFUL WEB SITES that will help you take your research further. Finally, the SET INDEX lists all the people and major topics covered in the complete set.

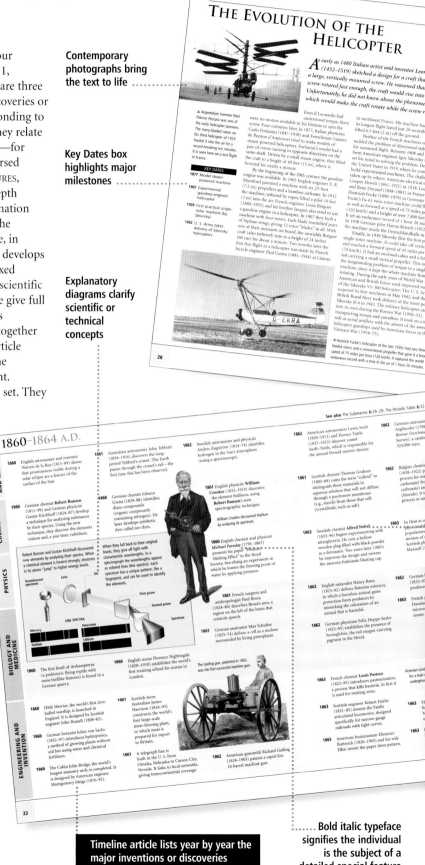

Contemporary photographs bring the text to life

Key Dates box highlights major milestones

Explanatory diagrams clarify scientific or technical concepts

Timeline article lists year by year the major inventions or discoveries

Bold italic typeface signifies the individual is the subject of a detailed special feature

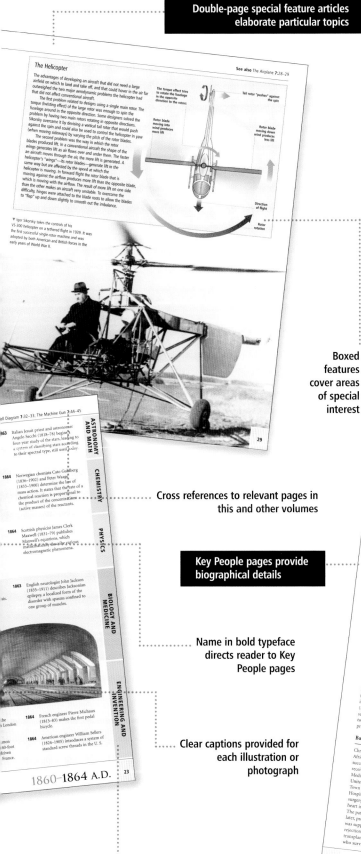

The Helicopter

See also *The Airplane* 7:28–29

The advantages of developing an aircraft that did not need a large airfield on which to land and take off, and that could hover in the air far outweighed the two major aerodynamic problems the helicopter had that did not affect conventional aircraft.

The first problem related to designs using a single main rotor. The torque (twisting effect) of the large rotor was enough to spin the fuselage around in the opposite direction. Some designers solved the problem by having two main rotors rotating in opposite directions. Sikorsky overcame it by devising a vertical tail rotor that would push against the spin and could also be used to control the helicopter in yaw (when moving sideways) by varying the pitch of the rotor blades.

The second problem was the way in which the rotor blades produced lift. In a conventional aircraft the shape of the wings generates lift as air flows over and under them. The faster an aircraft moves through the air, the more lift is generated. A helicopter's "wings"—its rotor blades—generate lift in the same way but are affected by the speed at which the helicopter is moving. As the forward flight the rotor blade that is moving against the airflow produces more lift than the opposite blade, which is moving with the airflow. The result of more lift on one side than the other makes an aircraft very unstable. To overcome this difficulty, hinges were attached to the blade roots to allow the blades to "flap" up and down slightly to smooth out the imbalance.

The torque effect tries to rotate the fuselage in the opposite direction to the rotors

Tail rotor "pushes" against the spin

Rotor blade moving into wind produces more lift

Rotor blade moving down wind produces less lift

Direction of flight

Rotor rotation

▼ Igor Sikorsky takes the controls of his VS-300 helicopter on a tethered flight in 1939. It was the first successful single-rotor machine and was adopted by both American and British forces in the early years of World War II.

29

The Scientific Revolution
1625–1774

In the early 1600s the idea of the Sun-centered solar system, put forward the previous century by Polish astronomer Nicolaus Copernicus, was championed by Italian scientist Galileo Galilei. This revolutionary idea was outside the teachings of the Roman Catholic church, and it brought Galileo into conflict with the officials of the notorious Inquisition, who sentenced him to house arrest for the last eight years of his life. English scientist and mathematician Isaac Newton (born in the year of Galileo's death) had no such problems with the church. However, he was involved in controversy with fellow scientists, such as Robert Hooke, whom Newton accused of stealing his ideas on several occasions.

But Galileo and Newton had one important thing in common: They were not just thinkers, they were also experimenters. Galileo developed his ideas about pendulums by using his pulse rate to time the swings of a chandelier in Pisa Cathedral. The story goes that he even tested his theory about falling weights by dropping cannonballs from the Leaning Tower of Pisa. Newton used a glass prism to split white sunlight into a spectrum when formulating his ideas about light. His law of universal gravitation was one of the bones of contention with Hooke, who claimed to have thought of it first. In America Benjamin Franklin experimented with lightning as well as inventing bifocal eyeglasses and a nonsmoking wood-burning stove.

At the beginning of the 18th century cheaper iron became available in quantity after Abraham Darby introduced a method of smelting iron using coke instead of charcoal. This innovation was to have important effects in the manufacturing industry—especially after James Watt developed his steam engine and paved the way for the Industrial Revolution that was to follow.

Bell Diagram 7:32–33; The Machine Gun 7:44–45

ASTRONOMY AND MATH

1863   Italian Jesuit priest and astronomer Angelo Secchi (1818–78) begins a four-year study of the stars, leading to a system of classifying stars according to their spectral type, still used today.

CHEMISTRY

1864   Norwegian chemists Cato Guldberg (1836–1902) and Peter Waage (1833–1900) determine the law of mass action. It states that the rate of a chemical reaction is proportional to the product of the concentrations (active masses) of the reactants.

PHYSICS

1864   Scottish physicist James Clerk Maxwell (1831–79) publishes Maxwell's equations, which mathematically describe various electromagnetic phenomena.

BIOLOGY AND MEDICINE

1863   English neurologist John Jackson (1835–1911) describes Jacksonian epilepsy, a localized form of the disorder with spasms confined to one group of muscles.

ENGINEERING AND INVENTION

1864   French engineer Pierre Michaux (1813–83) makes the first pedal bicycle.

1864   American engineer William Sellers (1824–1905) introduces a system of standard screw threads in the U.S.

1860–1864 A.D.   23

KEY PEOPLE 1950–1979 A.D.

People whose names appear in bold type have their own articles in this section or in the "Key People" section of another volume. Names in bold italics indicate they are the subject of a special feature.

Bardeen, John (1908–91)

John Bardeen was an American physicist. He was born in Madison, Wisconsin. He graduated in electrical engineering from the University of Wisconsin in 1928 and earned his doctorate in mathematical physics at Princeton in 1936. He worked as a junior fellow at Harvard before moving to the University of Minnesota in 1938. He worked at the Naval Ordnance Laboratory during World War II before joining Bell Laboratories in 1945. In 1951 he became professor of physics and electrical engineering at the University of Illinois, where he remained until 1975.

At Bell Labs Bardeen worked with **William Shockley** (1910–89) and **Walter Brattain** (1902–87), producing the point-contact transistor in 1947. Bardeen also collaborated with American physicists Leon Cooper (1930–) and John Schrieffer (1931–) at the University of Illinois, where in 1957 they formulated the BCS (Bardeen–Cooper–Schrieffer) theory of superconductivity. For his contributions to science Bardeen shared two Nobel prizes for physics.

Barnard, Christiaan (1922–2001)

Christiaan Neethling Barnard was a South African surgeon who carried out the first successful heart transplant operations. He received his medical training at Cape Town Medical School and was a researcher in the United States before returning to Cape Town to work at the Groote Schuur Hospital in 1958. He performed open-heart surgery and in 1967 transplanted a donor heart into 54-year-old Louis Washkansky. The patient died of pneumonia 18 days later, probably because his immune system was suppressed by drugs used to prevent rejection of the new heart. In 1968 Barnard transplanted a heart into Philip Blaiberg, who survived for 563 days.

Békésy, Georg von (1899–1972)

Georg von Békésy was a Hungarian-born American biophysicist who worked mainly on the science associated with the sense of hearing. He was born in Budapest and studied chemistry at Berne University, Switzerland, and after military service he studied physical optics in Budapest. He received his doctorate in 1923 and worked in the research laboratory of the Hungarian post office from 1924 to 1946. From 1939 to 1946 he was professor of experimental physics at Budapest University. He later went to the United States and was senior fellow of psychophysics at Harvard from 1949 to 1966. He then took up a similar appointment at the University of Hawaii, where he remained until he died.

Following his work on long-distance telephones, Békésy began studying the workings of the cochlea, the coiled organ in the inner ear that detects sound sensations. In 1960 he discovered how incoming sound waves stimulate the so-called organ of Corti, which in turn fires nerve impulses along the auditory nerve to the brain.

Bell Burnell, Jocelyn (1943–)

Susan Jocelyn Bell Burnell is a British radio astronomer who, with her research supervisor Antony Hewish (1924–), discovered the first pulsar. She was born in Belfast, Northern Ireland. She trained at Glasgow and Cambridge universities, receiving her doctorate in 1968. At Cambridge in 1967 Burnell and Hewish detected a regular 3.7-meter radio signal from outer space, which was the first detected pulsar (a rapidly rotating condensed neutron star), designated CP 1919. Over the following months three pulsars. Bell Burnell went on to discover the next three pulsars. She held a research fellowship at Southampton University, where she worked on gamma radiation. She went to work on X-ray astronomy at the Mullard Space Science Laboratory in London in 1974, and in 1982 joined the staff of the Royal Observatory, Edinburgh, to manage their James Clerk Maxwell telescope in Hawaii. She was professor of physics at the Open University, Milton Keynes, for ten

Black, James (1924–)

James Whyte Black is a Scottish biochemist who invented several new medicinal drugs that work by blocking actions in the body that cause unwanted symptoms. He graduated in medicine from St. Andrews University, Scotland, in 1946. He taught at various universities before joining Imperial Chemical Industries as a senior pharmacologist in 1958. He was appointed head of biological research at Smith, Kline, & French Laboratories in 1964 and moved to the Wellcome Research Laboratories in 1978. He became professor of analytical pharmacology at King's College Medical School, London, in 1984.

Black's first success was in 1962, when he prepared the beta-blocker nethalide, which relieves cardiac tension and is used in the treatment of angina, hypertension (high blood pressure), and tachycardia (abnormally rapid heart beat). It acts by blocking the stimulation of beta-receptors (nerve endings) in the sympathetic nervous system. He then developed a similar drug, propranolol. While at Smith, Kline, & French, in 1972 Black devised drugs that suppress the secretion of acids in the stomach—burimamide and cimetidine. They are used for the treatment of patients with gastric and duodenal ulcers.

Brattain, Walter (1902–87)

Walter Houser Brattain was an American physicist and a member of the team that invented the transistor. He was born in Amoy, China, and raised in Washington state. He was educated at the universities of Oregon and Minnesota, receiving his doctorate in 1929. He worked for Bell Laboratories as a research physicist until 1967 (with a break during World War II when he investigated magnetic submarine detection systems). After he retired, he taught at Whitman College, Washington, continuing research into phospholipads (large molecules that are a major part of

years and served as president of the Royal Astronomical Society from 2002 to 2004.

the outer membranes of cells).

Brattain studied the surface properties of semiconductors such as germanium, particularly their ability to rectify (change an alternating current, AC, into a direct current, DC). In 1947, with physicists **William Shockley** (1910–89) and **John Bardeen** (1908–91), he developed the point-contact transistor. The device acted as an amplifier and soon replaced the bulky vacuum tubes used in electronic circuits before that time.

Calvin, Melvin (1911–97)

Melvin Calvin was an American chemist who figured out the cycle of biochemical reactions in photosynthesis, the process by which green plants use the energy of sunlight to convert carbon dioxide and water into food. Calvin was born in St. Paul, Minnesota. He graduated in chemistry from the Michigan College of Mining and Technology in 1931, obtaining his doctorate from the University of Minnesota four years later. He then went to work with Hungarian-born English chemist Michael Polanyi (1891–1976) at Manchester University, England, where he became interested in chlorophyll. In 1937 he went to the University of California at Berkeley, and—except for a time on the Manhattan Project (involved in making the atom bomb) at Los Alamos during World War II—he remained there. In 1946 he became director of the Lawrence Radiation Laboratory at Berkeley, and in 1971 he became university professor of chemistry.

Calvin began working on photosynthesis in the mid-1940s using a green alga called *Chlorella*. He exposed it to radioactive carbon dioxide (labeled with carbon-14) in the dark and traced the radiation through various organic compounds before it arrived in the sugar glucose. Brief exposure to light moved the radioactivity onto phosphate compounds, some of which were then released to repeat the cycle.

Cockerell, Christopher (1910–99)

Christopher Sydney Cockerell was an English engineer who invented the hovercraft, also called an air-cushion vehicle (ACV). He graduated from Cambridge

Eckert, John

John Presper E computer engi late 1940s a some of the firs was born in Phil graduated in elec became a research joined by Americ

46

5

ASTRONOMY AND MATH

1627 German astronomer **Johannes Kepler** (1571–1630) publishes tables describing the motions of the planets; they become known as the *Rudolphine Tables* for the deceased Emperor Rudolph II, who funded their publication.

1631 English mathematician Thomas Harriot (1560–1621) introduces the symbols > (meaning "greater than") and < ("less than").

1631 French mathematician and astronomer **Pierre Gassendi** (1592–1655) makes the first observation of the transit of the planet Mercury across the Sun's disk.

1632 The first modern astronomical observatory is built at Leyden (now Leiden) in the Netherlands.

1633 Italian scientist *Galileo Galilei* (1564–1642) is condemned by the Roman Catholic Inquisition for refusing to withdraw his statement that the Sun—and not the Earth—is at the center of the Universe (which is against the teachings of the church at the time). Galileo is sentenced to house arrest in Florence.

1637 French mathematician **René Descartes** (1596–1650) introduces analytic geometry, a way of using equations in algebra to represent geometric lines and curves; it is also called coordinate geometry because it employs Cartesian (named after Descartes) coordinates to describe the positions of points.

1639 English astronomers William Crabtree (1610–44) and Jeremiah Horrocks (1617–41) make the first observation of the transit of the planet Venus across the Sun's disk.

1642 French scientist **Blaise Pascal** (1623–62) builds a wooden mechanical calculating machine. It is the first primitive step toward making a computer.

CHEMISTRY AND PHYSICS

1635 English mathematician and astronomer Henry Gellibrand (1597–1636) discovers magnetic declination, which is the angular difference between the directions of magnetic north (indicated by a compass) and true north.

1640 French mathematician Pierre de Fermat (1601–65) proposes Fermat's principle, which states that light always travels in straight lines.

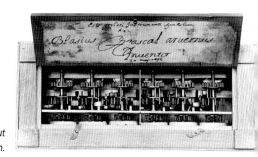

Pascal's calculating machine of 1642 could carry out simple functions of addition and subtraction.

BIOLOGY AND MEDICINE

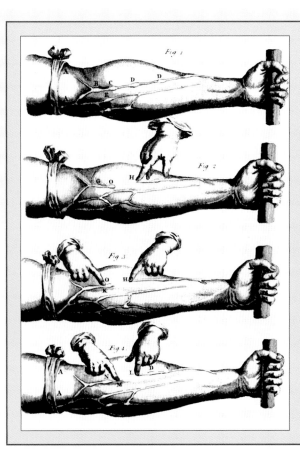

When studying animals, William Harvey began thinking about the circulation of blood around the body. He noticed that in a single hour the heart forces out far more blood than the total amount in the animal's body. The blood must, he reasoned, be going around in a continuous loop in what he called "a circle of ceaseless motion."

He announced his discovery in 1628 and published these diagrams to prove his point. He tied a ribbon around the upper arm—modern doctors call it a tourniquet—to make the veins stand out, with the valves showing up as swellings. They are one-way valves that allow the blood to flow back to the heart but not in the other direction. Harvey demonstrated that pressing one of the valves stops the blood flowing along a vein until the pressure is released.

1628 English physician **William Harvey** (1578–1657) explains the circulation of blood—the way in which blood is pumped by the heart through the lungs and around the rest of the body.

1630 Jesuit priests introduce to Europe the use of bark of the South American cinchona tree to treat malaria. Nearly 200 years later (in 1820) French chemists find that the bark contains the antimalaria drug quinine.

ENGINEERING AND INVENTION

1626 St. Peter's Basilica in Rome is finally consecrated, having taken 120 years to build; for over 250 years it stands as the world's largest Christian church.

1629 Italian engineer Giovanni Branca (1571–1640) makes a primitive steam turbine.

1631 French mathematician Pierre Vernier (1580–1637) invents the vernier measuring scale, which allows precise measurements to be made.

1644 French mathematician Marin Mersenne (1588–1648) discovers so-called Mersenne numbers, which are prime numbers of general form 2^n-1, where n is itself a prime. (A prime is a number that can be divided only by 1 and itself.)

A hydraulic press works according to the principles of Pascal's law of 1642. A large piston is connected to a small piston, and the setup is filled with an oily liquid. Because pressure is transmitted equally throughout a liquid, a small force on the small piston creates a greater force on the large piston. However, the larger piston will only move a fraction of the distance of the smaller one.

Small force   Large force

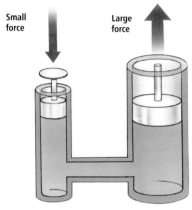

1644 Polish scientist and astronomer Johannes Hevelius (1611–87) observes that the planet Mercury shows phases, like the Moon (gradually changing from a crescent shape to a "half-Moon" to full Moon and back again).

Built in 1641, Hevelius's observatory in Danzig included the latest state-of-the-art astronomical instruments.

1642 French scientist **Blaise Pascal** (1623–62) puts forward Pascal's law, which states that the pressure within a liquid is the same everywhere. The principle underlies the working of all hydraulic machinery and the reason why it is possible to squeeze toothpaste from the end of a tube.

1644 German chemist Johann Glauber (1604–68) makes impure water glass (sodium silicate), a chemical later found to have many uses (including the preservation of eggs).

1641 Dutch naturalist Nicolaas Tulp (c.1593–c.1674) describes the first chimpanzee (*Pan troglodytes*) to be brought to the Netherlands. It was the first of the great apes to be discovered.

1641 German anatomist Franciscus Sylvius (1614–72) identifies the Sylvian fissure in the brain, which separates the temporal lobe (at the side) from the rest of the brain.

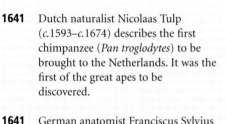
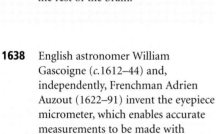

1642 German anatomist Johann Wirsung (1600–43) discovers the pancreatic duct, which carries digestive juices from the pancreas to the duodenum (part of the small intestine).

The chimpanzee was the first great ape to be discovered and was described by Dutch naturalist Nicolaas Tulp in 1641.

1638 English astronomer William Gascoigne (c.1612–44) and, independently, Frenchman Adrien Auzout (1622–91) invent the eyepiece micrometer, which enables accurate measurements to be made with telescopes and microscopes.

1639 A glassworks in Plymouth, Maryland, becomes one of the first "factories" in the American British colonies.

1640 French coachbuilder Nicolas Sauvage introduces the horse-drawn cab in Paris.

1641 Italian inventor Vincenzo Galilei (1606–49) tries (but fails) to make a pendulum clock, using a design produced by his father, **Galileo Galilei** (1564–1642).

1642 German engraver Ludwig von Siegen (1609–c.80) devises the mezzotint process, in which effects of light and shade are produced by scraping away parts of a roughened metal printing plate.

ASTRONOMY AND MATH

CHEMISTRY AND PHYSICS

BIOLOGY AND MEDICINE

ENGINEERING AND INVENTION

1625–1644 A.D.

GALILEO GALILEI (1564–1642)

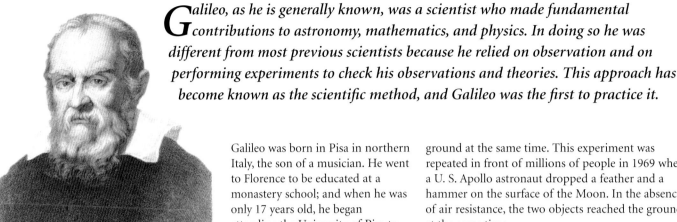

Galileo, as he is generally known, was a scientist who made fundamental contributions to astronomy, mathematics, and physics. In doing so he was different from most previous scientists because he relied on observation and on performing experiments to check his observations and theories. This approach has become known as the scientific method, and Galileo was the first to practice it.

▲ Galileo's genius made major contributions to astronomy, mathematics, and physics. But his single-minded refusal to compromise brought him into serious conflict with the church authorities.

Galileo was born in Pisa in northern Italy, the son of a musician. He went to Florence to be educated at a monastery school; and when he was only 17 years old, he began attending the University of Pisa to study medicine. He soon switched to mathematics and philosophy but left school in 1585 without a degree and began working as a teacher. His reputation spread, and in 1589 he became professor of mathematics at Pisa without any formal qualifications. He moved to Padua in 1592 and taught there until 1610.

"But it does move" (referring to the Earth orbiting the Sun, immediately after recanting to the Inquisition).

"The book of nature is written in that great book that ever lies before our eyes—I mean the Universe."

GALILEO

In 1582, while attending a service in Pisa Cathedral, Galileo noticed the regular movements of a lamp swinging in the air draft above his head. He made his own simple pendulum—for that is what the swinging lamp was—consisting of a weight at the end of a length of string, and then timed its swings. In those days stopwatches did not exist, and he used the beat of his own pulse to time the swings. He found that the time of each swing depended only on the length of the pendulum and was independent of the size of the weight. He suggested that a pendulum could be used to measure time, and pendulums were, in fact, used later to regulate mechanical clocks.

Since the time of the Greek philosopher *Aristotle* (384–322 B.C.) people thought that the speed at which an object falls depends on its weight. As an experimental scientist Galileo put this idea to the test in about 1602. He dropped two cannon balls of different weight from the Leaning Tower of Pisa— according to tradition—and showed that they hit the

ground at the same time. This experiment was repeated in front of millions of people in 1969 when a U. S. Apollo astronaut dropped a feather and a hammer on the surface of the Moon. In the absence of air resistance, the two objects reached the ground at the same time.

The development of the telescope in the early 1600s stimulated Galileo's interest in astronomy. He made telescopes of his own—ever the practical scientist—and pointed them at the Moon. He made drawings recording the Moon's mountains and craters at different times during the month. He then studied Jupiter and in 1610 announced that the planet had four moons of its own. A year later he aimed his telescope at the Sun and noted that sometimes small black spots moved slowly across the Sun's disk. The apparent movement of sunspots also convinced him that the Sun is rotating slowly on its axis.

Galileo's discoveries about the Sun and planets finally convinced him that the Sun is at the center of the Universe (or solar system, as we would call it), a theory that had first been published in 1543 by Polish astronomer *Nicolaus Copernicus* (1473–1543). But the idea went against the long-held teachings of

KEY DATES

1582 Constancy of pendulum's swing

1602 Law of falling bodies

1610 Observes four of Jupiter's moons

1611 First views sunspots

1633 Condemned by the Inquisition

Falling Objects

According to the teachings of the ancient Greek philosopher Aristotle, a heavy object, such as a large rock, must fall faster than a light one, such as a pebble. This assumption was based on logic and reflected the logical approach of the Greek philosophers. Galileo was not so sure. He put the theory to the test. The story goes that he dropped a pair of cannon balls of unequal size from the top of the Leaning Tower of Pisa. He showed that the two fell at exactly the same speed because they both hit the ground at the same time. He also tried to demonstrate that they fell at a constant acceleration. (Acceleration is the rate at which speed changes.) The illustration shows the results he would have obtained had he been able to measure acceleration. As an object falls farther, it moves faster, but the acceleration is constant—32 feet (9.8 m) per second per second. This is now known as the acceleration of free fall, or the acceleration due to gravity.

▲ Galileo gained fame by demonstrating his telescope to the members of Italian high society.

Object falls from here

16 ft (4.9 m)

0 ft (m) per second

After 1 second

32 ft (9.8 m) per second

48 ft (14.7 m)

After 2 seconds

64 ft (19.6 m) per second

80 ft (24.5 m)

96 ft (29.4 m) per second

After 3 seconds

Aristotle, who postulated an Earth-centered Universe. What was worse, the idea was contrary to the teachings of the Roman Catholic church. Church officials asked Galileo not to spread the idea, but in 1632 he published it in a book. Sales of the book were banned. In 1633 he was arrested by the Inquisition, taken to Rome, and under threat of torture, made to recant his views.

Old and in failing health, Galileo was banished to his home near Florence. He went blind in 1637, possibly because of eye damage caused by looking at the Sun. He was not allowed to move around freely, although he was allowed occasional visitors. One of them was English poet John Milton (1608–74). Coincidentally, Milton also went blind—one of his best-known sonnets was "On His Blindness," published in 1645. Galileo's own failing eyesight made him bitter because he could no longer observe the Universe that had been his life's work. When he died in 1642, he was suffering from arthritis and high blood pressure. Pope Urban VIII refused to forget his feud with Galileo, and he was buried unceremoniously in the Church of Santo Croce in Florence.

◄ Galileo made his own diagram of the Sun-centered Universe proposed by Copernicus. He also included his own discovery: the four moons of Jupiter (shown second from top).

ASTRONOMY AND MATH

1645 Flemish cartographer Michael Langrenus (1600–75) publishes the first map of the Moon.

1650 Italian astronomer Giovanni Riccioli (1598–1671) identifies the first binary (double) star, Mizar, in the constellation Ursa Major (Great Bear).

The pyramid of numbers created by Blaise Pascal is easy to construct. Just start with 1. Then on the next row write two 1s (all rows start and end with 1). On row three after the initial 1 write the number that is the sum of the two immediately above it: 1 + 1 = 2. Now do it again: Start with 1 and then write in numbers that are the sum of the two above them (1 + 2 = 3). And so on and so on. Pascal's triangle has important applications in algebra.

1653 French scientist **Blaise Pascal** (1623–62) devises Pascal's triangle of numbers, in which each number is the sum of the two numbers above it. It is later to have important applications in mathematics, e.g., binomial theorem.

Right: By developing a new technique to polish the lenses of his homemade telescope, Christiaan Huygens was the first person to observe Saturn's rings in 1655.

CHEMISTRY AND PHYSICS

1645 Italian physicist **Evangelista Torricelli** (1608–47) constructs the first mercury barometer. He also discovers the Torricellian vacuum, produced by inverting a tube full of mercury into a dish (the vacuum forms in the space at the closed top end of the inverted tube).

1646 English scientist Thomas Browne (1605–82) coins the word "electricity" (which at that time was limited to static electricity).

1646 French scientist **Blaise Pascal** (1623–62) demonstrates the existence of atmospheric pressure and confirms that it varies with altitude.

BIOLOGY AND MEDICINE

1645 English physician Daniel Whistler (1619–84) first diagnoses the childhood disease rickets. It is later (1651) also independently described by his compatriot Francis Glisson (1597–1677).

1647 French anatomist Jean Pequet (1622–74) describes the thoracic duct in animals. This duct carries the fluid lymph from the lower body to a vein in the neck. In 1653 Swedish scientist Olof Rudbeck (1630–1702) discovers the lymphatic vessels in humans.

1649 English physician Henry Power (1623–68) discovers the extremely narrow capillary blood vessels.

1658 Dutch naturalist Jan Swammerdam (1637–80) describes red blood cells (erythrocytes).

ENGINEERING AND INVENTION

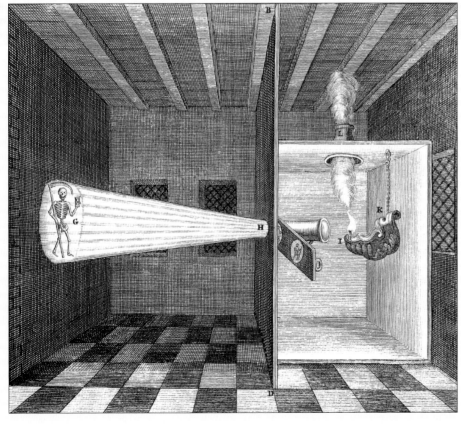

1646 German scholar and inventor Athanasius Kircher (1601–80) describes a magic lantern that projects hand-drawn images.

1647 French scientist **Blaise Pascal** (1623–62) invents a primitive version of the roulette wheel to test his ideas on probability.

1650 German inventor Stephen Farfler constructs a three-wheeled invalid chair.

1652 Dutch hydraulic engineer Cornelius Vermuyden (1595–c.1683) completes the drainage of a large area of the Fens—the low marshy lands in the east of England— making them suitable for growing farm crops.

The magic lantern described by Athanasius Kircher used "slides" made of images painted on a mirror.

1649 French mathematician and philosopher **Pierre Gassendi** (1592–1655) publishes *Syntagma Philosophiae Epicuri*, a study of Greek philosopher Epicurus (*c.*342–270 B.C.), who holds that all matter is made up of atoms.

1659 English physician Thomas Willis (1621–75) first describes typhoid fever.

1659 Italian anatomist Marcello Malpighi (1628–94) discovers the lymph nodes, enlarged structures where lymph vessels come together. Two years later he confirms the existence of capillary blood vessels.

1654 Polish scientist Johannes Hevelius (1611–87) makes a closed-ended thermometer (previous versions of liquid-in-glass thermometers had one end open and had to be very long).

1654 German physicist Otto von Guericke (1602–86) of Magdeburg invents an air pump (which is the name given at that time to a vacuum pump). He gives a demonstration and removes the air from between two copper hemispheres. Two teams of horses fail to pull it apart. The hemispheres, which become known as the Magdeburg spheres, are held together by atmospheric pressure.

1656 Dutch scientist **Christiaan Huygens** (1629–95) designs a pendulum clock. His chronometer for use at sea, which he invents in 1659, fails to keep accurate time.

1655 Dutch scientist **Christiaan Huygens** (1629–95) uses a homemade telescope to observe the rings around the planet Saturn. He also discovers Titan, the largest of Saturn's moons.

1656 English mathematician John Wallis (1616–1703) introduces the infinity sign ∞. He first publishes it in his book *Arithmetica Infinitorum*.

1650 German scholar and inventor Athanasius Kircher (1601–80) demonstrates that sound will not travel in a vacuum.

1657 Dutch scientist **Christiaan Huygens** (1629–95) writes the first book on mathematical probability theory.

1659 German mathematician Johann Rahn (1622–76) introduces the division sign (÷) to mathematics.

1650 The first properly equipped chemistry laboratory is established at the University of Leyden (now Leiden) in the Netherlands.

The lymphatic system is a collection of vessels that carry the watery fluid lymph from the tissues into the bloodstream and to the heart. The vessels pass through oval structures called lymph nodes (dark dots on the diagram, right) that were first discovered by Marcello Malpighi in 1659. Scientists now know that lymph nodes contain special white blood cells (lymphocytes) that produce antibodies to fight infection.

LYMPHATIC SYSTEM

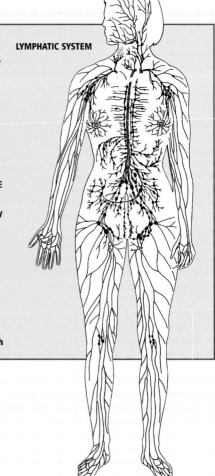

LYMPH NODE

Lymph vessels

Blood supply

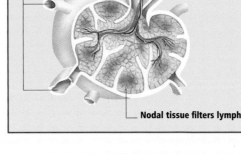

Nodal tissue filters lymph

1658 English scientist **Robert Hooke** (1635–1703) makes a watch regulated by a hairspring (the power from the main spring of the watch is released very gradually, controlled by the oscillations of the hairspring).

1645–1659 A.D.

THE PENDULUM CLOCK

The Chinese constructed the first mechanical clocks, which were in effect large, slowly turning waterwheels made of wood. Metal clocks powered by falling weights date from the 1300s, but they were unreliable and inaccurate. What was needed was a way of closely regulating the mechanism, and that arrived in the 1600s as the first practical application of the pendulum.

▲ Galileo's pupil Vincenzo Viviani (1622–1703) drew the above sketch of a pendulum clock, shown as a reconstruction (right).

In 1582 Italian scientist **Galileo Galilei** (1564–1642) demonstrated that a pendulum always swings at a constant rate. He also proved that the rate of swing depends only on the length of the pendulum and not on the size of the weight swinging at its end. In mathematical terms the time taken for one swing is proportional to the square root of the length of the pendulum. For example, a pendulum 39 inches (99 cm) long takes one second to make one swing (forward and back). So, if a pendulum of that length keeps swinging, it can mark off seconds of time.

This idea occurred to Galileo, and in 1641—a year before he died—he instructed his son Vincenzo (1606–49) how to make a clock regulated by a pendulum. Vincenzo did not complete the job, and it was not until 1657 that the first pendulum clock appeared. It was designed by the Dutch scientist **Christiaan Huygens** (1629–95) a year before and constructed by clockmaker Salomon Coster in The Hague. It kept time to within five minutes a day and was much more accurate than any earlier clocks.

Clock pendulums use a metal rod rather than a piece of string. However, the metal of the rod does not stay at a constant length—a factor that is crucial for accuracy—but varies depending on temperature. The rod gets longer when it is warm and shorter when it is cold. A clock with a one-second pendulum,

for instance, needs an increase in length of just 0.0009 inches (0.025 mm) in order to lose about one second a day, and a steel rod expands by that much with a temperature rise of only 4°F (2°C).

To overcome the problem, inventors soon devised various ways of making a pendulum that kept a constant length. In 1722 English inventor George Graham (1673–1751) designed the mercury pendulum (announced in 1726), which has a glass jar of mercury as the pendulum's weight. When the pendulum expands downward because of a rise in temperature, the change is counterbalanced by the upward expansion of the mercury in the jar. Another solution, the gridiron pendulum, was invented by English clockmaker John Harrison (1693–1776) in 1728. His design has a grid of alternate brass and steel rods. Brass expands more than steel, so the expansion of the brass compensates for the lesser expansion of steel. A pendulum rod made of concentric tubes of iron and zinc achieves a similar result. Today pendulum rods are made from invar, an alloy of iron and nickel that expands very little when heated. It is also used for making measuring tapes and tuning forks, in which stable dimensions are important.

Have you ever wondered why a grandfather clock (known technically as a long-case clock) is so tall? It has to be long enough to house a 39-inch (99-cm) one-second pendulum, common to all such clocks. They repeatedly go tick, tock every second.

The Constant Pendulum

Galileo was the first person to notice that a pendulum has a constant rate of swing. Observing a swinging lamp in Pisa Cathedral and using his own pulse to keep time, he found that the rate of swing depends only on the length of the pendulum—it is independent of the size of the weight on the end and the angle of swing (as long as the angle is fairly small).

In modern mathematical terms the time of the swing (called its period) is proportional to the square root of the length of the pendulum. So, for a long period a very long pendulum is needed. A one-second pendulum has to be 39 inches (99 cm) long.

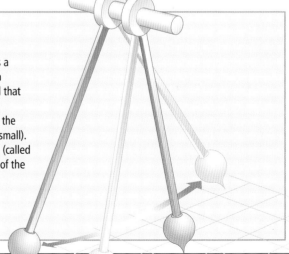

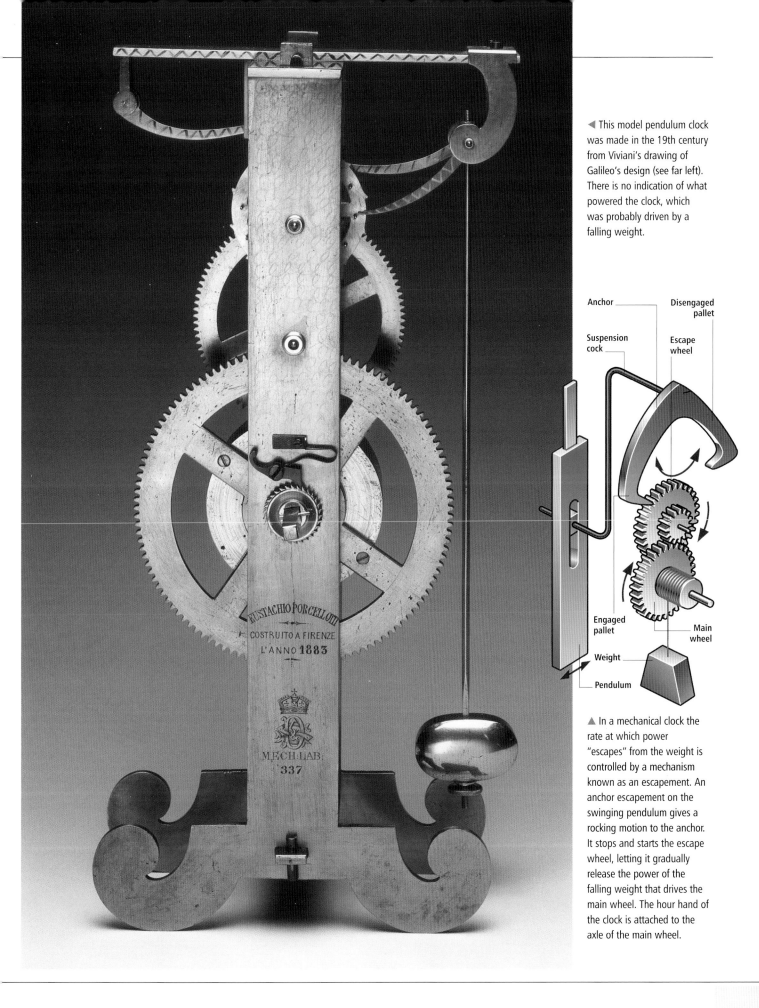

◀ This model pendulum clock was made in the 19th century from Viviani's drawing of Galileo's design (see far left). There is no indication of what powered the clock, which was probably driven by a falling weight.

Anchor — Disengaged pallet

Suspension cock — Escape wheel

Engaged pallet — Main wheel

Weight

Pendulum

▲ In a mechanical clock the rate at which power "escapes" from the weight is controlled by a mechanism known as an escapement. An anchor escapement on the swinging pendulum gives a rocking motion to the anchor. It stops and starts the escape wheel, letting it gradually release the power of the falling weight that drives the main wheel. The hour hand of the clock is attached to the axle of the main wheel.

EUSTACHIO PORCELLOTTI
COSTRUITO A FIRENZE
L'ANNO 1883

MECH.LAB.
337

ASTRONOMY AND MATH

1664 English scientist **Robert Hooke** (1635–1703) describes Jupiter's Great Red Spot (a major feature of Jupiter's surface now known to be due to a gigantic storm in the atmosphere). In the same year Hooke proposes that the planets are held in their orbits by the force of gravity between them and the Sun.

Jupiter's Red Spot, discovered by Robert Hooke in 1664, is now known to be a huge permanent storm in the planet's upper atmosphere.

CHEMISTRY AND PHYSICS

1661 Irish scientist **Robert Boyle** (1627–91) publishes his book *The Sceptical Chymist*, in which he defines chemical elements for the first time. A year later he formulates Boyle's law, which states that at a fixed temperature the pressure of a gas is inversely proportional to its volume.

Robert Boyle was a scientist who was accomplished in many fields but is best remembered for his work on gases.

1665 English scientist **Robert Hooke** (1635–1703) proposes the wave theory of light, but his proposal is largely ignored until championed by **Christiaan Huygens** in 1678.

1668 English mathematician John Wallis (1616–1703) proposes that in a collision total momentum does not change. This is the law of conservation of momentum: The momentum (mass x velocity) of objects before a collision is the same as their combined momentum afterward.

BIOLOGY AND MEDICINE

1660 French physicist Edmé Mariotte (1620–84) discovers the blind spot on the retina of the eye (at the place where the optic nerve joins the retina).

1665 English scientist **Robert Hooke** (1635–1703) coins the word "cell" to describe the "little boxes" he observes in plant tissues using a compound microscope of his own invention.

1665 English physician Richard Lower (1631–91) performs the first blood transfusion, from a dog to another dog. Two years later in France physician Jean Denis (1625–1704) performs a transfusion from a sheep to a human. (The patient died.)

1666 English physician Thomas Sydenham (1624–89) pioneers the use of iron compounds in treating anemia.

1669 Dutch naturalist Jan Swammerdam (1637–80) describes metamorphosis in insects—the sequence of changes from egg to larva to pupa to adult or imago (as in butterflies).

1670 English physician Thomas Willis (1621–75) detects the presence of sugar in the urine of patients with diabetes.

ENGINEERING AND INVENTION

1660 English scientist **Robert Hooke** (1635–1703) devises the anchor escapement to regulate a pendulum-driven clock.

1661 Dutch scientist **Christiaan Huygens** (1629–95) invents the manometer, a device for measuring gas pressure.

1663 Scottish mathematician and inventor **James Gregory** (1638–75) proposes a design for a reflecting telescope.

1664 Italian engineer Giuseppe Campani (1635–1715) creates a lens-grinding lathe (for making lenses for optical instruments).

1666 French engineer Jean de Thévenot (1620–92) makes the spirit level.

1667 English scientist **Robert Hooke** (1635–1703) invents the anemometer, an instrument for measuring wind speed. It is later to have important applications in the study of weather.

1668 English scientist and mathematician *Isaac Newton* (1642–1727) builds a reflecting telescope.

1670 French winemaker Dom Pérignon (1638–1715) creates champagne.

1674 English glassmaker George Ravenscroft (1618–81) develops lead crystal glass.

1674 Dutch military engineer Menno von Coehoorn (1641–1704) invents the trench mortar.

See also The Pendulum Clock **4:**12–13; Telescopes **4:**24–25

ASTRONOMY AND MATH

1665 English scientist and mathematician *Isaac Newton* (1642–1727) introduces the binomial theorem—a key theorem in algebra that allows the expansion of the expression $(x + y)^n$. This was only one of many advances in mathematics made by Newton at about this time.

1666 Italian astronomer **Giovanni Cassini** (1625–1712) observes the polar ice caps on Mars. (We now know that the "ice" is mainly frozen carbon dioxide.)

1667 French astronomer Jean Picard (1620–82) develops his own version of a micrometer eyepiece for making accurate star measurements with an astronomical telescope.

1670 French astronomer Jean Picard (1620–82) measures the length of part of the meridian (a line of longitude), thus allowing accurate calculation of the circumference of the Earth.

1671 Italian astronomer **Giovanni Cassini** (1625–1712) discovers Iapetus, Saturn's third-largest moon (after Titan and Rhea).

CHEMISTRY AND PHYSICS

1669 Danish scientist Erasmus Bartholin (1625–98) discovers double refraction—when an object is viewed through certain crystals (such as Iceland spar), two images are seen. The phenomenon remained a mystery until it was explained in 1808 by French physicist Étienne Malus (1775–1812).

1669 German alchemist Hennig Brand (*c.*1630–*c.*92) discovers phosphorus (in urine). This was the first discovery of a new element since prehistoric times.

1669 German chemist Johann Becher (1635–82) proposes the (incorrect) phlogiston theory of combustion (which states that materials get lighter in weight by releasing "phlogiston" when they burn).

1671 English scientist and mathematician *Isaac Newton* (1642–1727) demonstrates that a glass prism splits white light into a spectrum of rainbow colors—the phenomenon known as dispersion of light.

Isaac Newton first used a prism in 1671 to split white light into the colors of the rainbow.

BIOLOGY AND MEDICINE

1672 Dutch naturalist Jan Swammerdam (1637–80) identifies the human ovaries. In the same year his compatriot, anatomist Regnier de Graaf (1641–73), uncovers egg-containing follicles (known as Graafian follicles) within the ovaries.

1673 Dutch scientist **Antonie van Leeuwenhoek** (1632–1723) begins writing letters to the newly formed Royal Society of London describing what he has observed under a microscope.

ENGINEERING AND INVENTION

Dom Pierre Pérignon ("Dom" is a title given by the Roman Catholic Church) was a French Benedictine monk who developed the process for making champagne wine. People in southern Europe have made wine since the time of the ancient Greeks and Romans, and the district along the valley of the Marne River became famous for its chardonnay grapes that result in a fine white wine. To make champagne, the winemakers allow chardonnay to ferment in large vats. Then they usually blend it with wines from previous years and add sugar and yeast before putting it in bottles made from thick, dark-green glass. The wine ferments for a second time in the bottles, and the gas (carbon dioxide) produced results in an effervescent wine known as champagne and appropriately nicknamed "bubbly." Using the same process, nonblended wine produces "vintage" champagne. Dom Pérignon was born Pierre Pérignon in Saint-Menehould, son of a judge's clerk. He entered the Benedictine order when he was 19, and within ten years he became cellarmaster at the Abbey of Hautvillers. Today the Dom Pérignon name is owned by the Moët & Chandon company.

1660–1674 A.D.

THE BAROMETER AND VACUUMS

We know that air has mass and that atmospheric pressure results from the weight of the atmosphere pressing down on every object on Earth. But these statements have not always been taken as fact. In the 1640s an Italian scientist set out to measure air pressure; in doing so, he proved the existence of the vacuum and invented the barometer.

▲ During his experiments on air pressure Evangelista Torricelli produced the first vacuum. His findings led to the invention of the mercury barometer and proved once and for all that vacuums do indeed exist.

Evangelista Torricelli (1608–47) trained as a mathematician and in 1641 went to work as an assistant to the aged **Galileo** (1564–1642), who always maintained that there is no such thing as a vacuum. In 1645, aided by his own assistant Vincenzo Viviani (1622–1703), Torricelli took a 6.6-foot (2-m) glass tube, sealed at one end, and filled it with mercury. Keeping his thumb over the open end, he upended the tube in a dish full of mercury and then removed his thumb. Some of the mercury ran out into the dish, and the mercury level in the tube dropped to about 30 inches (76 cm). But what was preventing it all from escaping?

Torricelli reasoned that the weight of the atmosphere (air) pressing on the surface of the mercury in the dish equaled the weight of the mercury left in the tube. The height of the mercury column is therefore a measure of air pressure. The whole device is known as a barometer. Toricelli also noticed that the height of the column varied slightly from day to day depending on the weather and deduced that atmospheric pressure must also vary daily. In 1647 French mathematician **René Descartes** (1596–1650) added a vertical scale to a Torricelli barometer and used it to record weather observations. To this day, air pressure is one of the most important factors considered by weather forecasters. Pressure is still sometimes measured in inches or millimeters of mercury— "normal" pressure is 30 inches (760 mm).

Atmospheric pressure also varies with altitude. The pressure at the top of a mountain is less than that at the foot of the mountain. (The pressure outside a high-flying jet plane is practically zero.) In 1771, over a century after Torricelli's death, Swiss geologist Jean Deluc (1727–1817) began using a sensitive barometer to measure the heights of mountains. A modern altimeter, used in airplanes, is also a modified barometer, although not the mercury type.

Torricelli's barometer was not very portable, even if Jean Deluc carried one up mountains. In 1797 a French scientist, Nicolas Fortin (1750–1831), invented a portable mercury barometer. The mercury reservoir was contained in a leather bag. By turning a screw, the bag could be squeezed slightly to bring the surface of the mercury in line with a level indicated by a pointer,

and a vernier scale at the top of the tube allowed very accurate pressure readings to be taken.

Returning to Torricelli's experiment, what was in the space above the mercury at the top of the closed tube? The short answer is "nothing." In fact, it was a vacuum, a space in which there is nothing at all. Scientists soon wanted to study vacuums and their effects, and needed a way to produce them in the laboratory. In 1654 physicist Otto von Guericke (1602–86), mayor of Magdeburg, Germany, invented the first air pump, so-called because it was designed to pump air out of a vessel. (Today we would call it a vacuum pump.) He used it to remove the air from

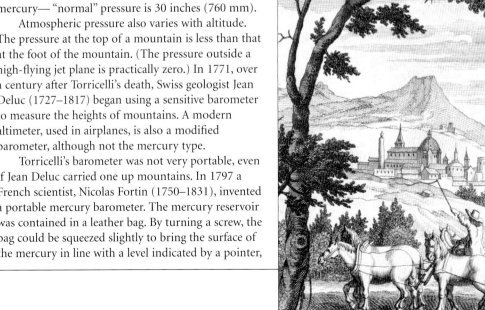

between a pair of copper hemispheres, which became held tightly together by the force of atmospheric pressure. Even 16 horses could not pull them apart. Von Guericke's demonstration apparatus became known as the Magdeburg spheres.

Other, more efficient vacuum pumps followed. In 1703 English physicist Francis Hawksbee (*c.*1666–1713) made a vacuum pump that Irish scientist **Robert Boyle** (1627–91) employed in his studies of air and other gases. German physicist Heinrich Geissler (1815–79) used his pump of 1855 to study electrical discharges at low pressures, and ten years later German-born British scientist Hermann Sprengel (1834–1906) mechanized Geissler's pump so that it could produce high vacuums. Still used today, it is a mercury vapor pump (or diffusion pump) in which the vapor "captures" molecules of air and carries them away to produce a vacuum. It had very important applications in science, leading to the finding of the rare gases in air, the discovery of the electron, and the invention of the electric light bulb, among other things.

The Mercury Barometer

To recreate Torricelli's original experiment, the closed glass tube is completely filled with mercury and then placed upside down in the dish of mercury. The mercury column in the tube falls slightly, leaving a vacuum in the space above. Atmospheric pressure acting on the surface of the mercury in the dish holds up the mercury column. The height of the column above the surface measures the atmospheric pressure.

Vacuum

Atmospheric pressure measured in inches/mm of mercury

Mercury

Atmospheric pressure

Mercury

▼ Otto von Guericke's original air pump used a long lever to work a piston up and down in a vertical cylinder.

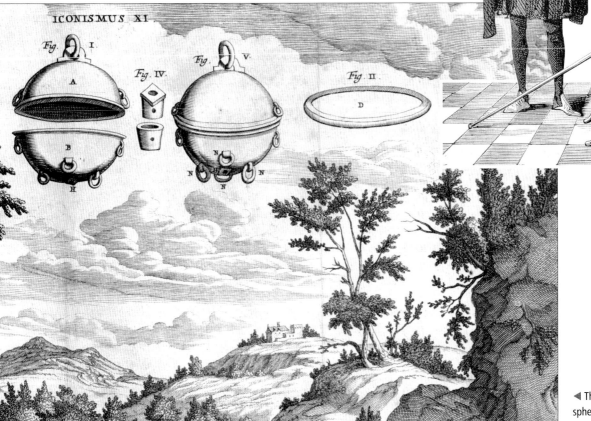

ICONISMUS XI

◀ The famous Magdeburg spheres were, in fact, two hemispheres of copper. After the air had been pumped out, creating a vacuum, 16 horses could not pull them apart.

ASTRONOMY AND MATH

1675 Italian astronomer **Giovanni Cassini** (1625–1712) discovers the major gap in Saturn's rings, now known as the Cassini division.

1675 The Royal Greenwich Observatory is completed near London and gives its name to the Greenwich Meridian (at longitude 0°).

In this picture of Greenwich Observatory the astronomer on the left uses a quadrant, while the one on the right looks through a long telescope.

CHEMISTRY AND PHYSICS

1675 English scientist and mathematician *Isaac Newton* (1642–1727) proposes the corpuscular theory of light: that light travels as a series of rapidly moving minute particles. He does not publish his theory until 1704.

1676 French physicist Edmé Mariotte (1620–84) proposes Mariotte's law, which is identical to Boyle's law of 1662: that at a fixed temperature the pressure of a gas is inversely proportional to its volume.

1676 English scientist **Robert Hooke** (1635–1703) proposes Hooke's law, which states that when an elastic object stretches, the stress (force per unit area) in it is proportional to the strain (change in dimensions). Hooke sums it up in Latin as *Ut tensio, sic vis.*

1676 Danish astronomer Ole Rømer (1644–1710) measures (inaccurately) the speed of light.

1678 Dutch scientist **Christiaan Huygens** (1629–95) takes up the wave theory of light proposed by English scientist **Robert Hooke** (1635–1703) in 1665: that light travels as a series of minute waves.

BIOLOGY AND MEDICINE

1676 Dutch scientist **Antonie van Leeuwenhoek** (1632–1723) reports his observations of bacteria, using a simple microscope made with lenses he ground himself. A year later he observes human sperm.

1681 On the Indian Ocean island of Mauritius the dodo (*Raphus cucullatus*), a large flightless bird of the pigeon family, becomes extinct.

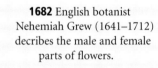

1682 English botanist Nehemiah Grew (1641–1712) decribes the male and female parts of flowers.

1683 English physician Thomas Sydenham (1624–89) gives the first full description of gout.

The dodo weighed about 50 pounds (23 kg) and was bigger than a turkey.

ENGINEERING AND INVENTION

1675 Dutch scientist **Christiaan Huygens** (1629–95) invents the oscillating balance and hairspring regulator for clocks.

1675 English watchmaker Thomas Tompion (1639–1713) makes a watch with a hairspring escapement.

1675 Irish scientist **Robert Boyle** (1627–91) devises a hydrometer (an instrument for measuring the relative density of a liquid). It soon finds important applications in science and industry.

1676 English scientist **Robert Hooke** (1635–1703) invents the universal joint (for connecting two driven shafts joined at an angle).

1679 German alchemist and glassmaker Johann Kunckel (*c.*1630–*c.*1702) creates a type of red glass that is used for making artificial (fake) rubies.

1679 French physicist **Denis Papin** (1647–1712) produces a steam digester, the forerunner of the pressure cooker.

1680 English clockmaker Daniel Quare (1648–1724) makes a repeating watch that chimes the hours and repeats the chime when a button is pressed.

This ornate watch by Thomas Tompion dates from c.1700–10.

1679 German mathematician **Gottfried Leibniz** (1646–1716) introduces binary arithmetic, which uses only two digits. Today it is employed by all computers.

1682 English astronomer **Edmond Halley** (1656–1742) plots the course of Halley's comet. In 1705 he correctly predicts that it will return in 1758.

1679 English scientist **Robert Hooke** (1635–1703) proposes the inverse square law of gravity.

1687 English scientist and mathematician *Isaac Newton* (1642–1727) publishes his major work *Principia*, in which he sets out his various theories in astronomy, mathematics, and physics.

1683 The wild boar (*Sus scrofa*) becomes extinct in the British Isles.

1686 English naturalist John Ray (1627–1705) proposes the word "species" to describe an interbreeding group of plants.

1683 Japanese mathematician Seki Kowa (1642–1708) introduces the use of determinants into mathematics. A determinant is a square array of numbers (elements) useful in solving simultaneous equations and other mathematical problems.

1684 Italian astronomer **Giovanni Cassini** (1625–1712) discovers Dione and Thetys, two of Saturn's moons.

1687 German astronomer Gottfried Kirch (1639–1710) discovers that Zeta Cygni is a variable star.

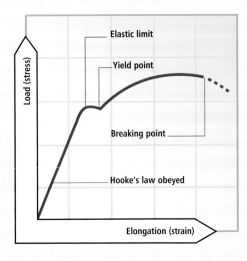

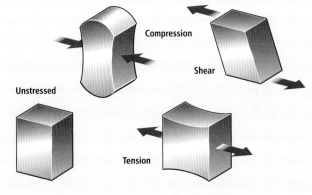

Left: The graph shows that as a piece of metal is stretched, the load applied (stress) is at first proportional to the elongation (strain). In this phase the metal obeys Hooke's law of 1676, which states that stress is proportional to strain. But beyond the elastic limit the metal is permanently stretched and cannot return to its original length. At the yield point the metal yields and stretches rapidly with very little extra load until it breaks.

Right: The three main kinds of stress. Compression takes place when a material is squeezed, shear results when the top and bottom are pulled in opposite directions, and tension is caused by a stretching force.

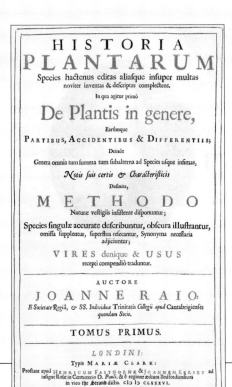

John Ray published his three-volume history of plants, Historia Plantarum, between 1686 and 1704. He devised a system of plant classification and introduced the term "species" to describe members of an interbreeding group. His writings on botany earned him the name "Father of Natural History."

1680 Irish scientist **Robert Boyle** (1627–91) invents a match that uses a mixture of sulfur and phosphorus. Smelly and poisonous, it remained the basic "strike anywhere" match for more than 200 years.

1687 French physicist Guillaume Amontons (1663–1705) invents a hygrometer (an instrument for measuring the humidity of the atmosphere).

1689 German maker of woodwind instruments Johann Denner (1655–1707) develops the clarinet (as a single-reed instrument with no keys).

ASTRONOMY AND MATH

CHEMISTRY AND PHYSICS

BIOLOGY AND MEDICINE

ENGINEERING AND INVENTION

1675–1689 A.D.

ISAAC NEWTON (1642–1727)

*O*ne of the first scientists to truly merit the title of genius, Isaac Newton made fundamental breakthroughs in mathematics and established the basic laws that became the cornerstones of astronomy and physics. The first scientist to receive a knighthood, his name lives on as the newton, the modern SI unit of force.

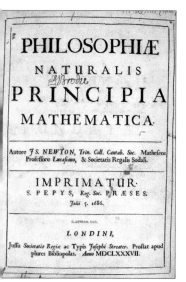

▲ Isaac Newton's *Principia* was one of the most important books on science ever written. It deals with theories in astronomy, mathematics, and physics.

Isaac Newton was born in eastern England. He was brought up by his grandmother and educated at a local school before going to Trinity College (Cambridge University). He received his bachelor's degree in 1665 and was forced to remain in the countryside because of the plague that raged in London at that time. At first he concentrated on mathematics, working out the principles of "fluxions," which were to lead to differential calculus.

In 1667 Newton received a fellowship of Trinity College and became professor of mathematics in 1669. He turned his attention to what happens when objects move—what makes them start moving and what stops them. His conclusions are summed up in Newton's three laws of motion. All of the laws can be observed by watching a game of pool, although you do not have to be a physicist to play it!

Newton's next contribution was to have a profound effect on astronomy. According to the well-known story, he was sitting in an orchard when he saw an apple fall. Why did it fall? Newton concluded that it was attracted toward the Earth by a force, which we now call the force of gravity. He also deduced that every object behaves as if its mass were concentrated in one place, its center of gravity (now called the center of mass). Applying his own laws of motion, he figured out that all objects in the Universe are affected by such gravitational forces—it is gravity that keeps the Moon in orbit around the Earth and the Earth in orbit around the Sun. He produced a formula, the universal law of gravitation, that states the gravitational force between any two objects—two pool balls or even two stars—is equal to the product of their masses and inversely proportional to the distance between them.

"If I have seen further it is by standing on the shoulders of giants."

"O Diamond! Diamond! Thou little knowest the mischief done!" (to a dog that overturned a candle and set fire to some papers that had taken years to write).

ISAAC NEWTON

The English scientist **Robert Hooke** (1635–1703) also devised a law of gravity in about 1678 and published his ideas a few years later. This led to a bitter dispute between the two great men.

KEY DATES

1665 Binomial theorem (math)

c.1665 Law of gravity and center of gravity

1668 Newton's reflecting telescope

1671 White light into a spectrum

1675 Corpuscular theory of light

1687 Book *Principia* published

1704 Book *Optics* published

Newton's Laws

Newton formulated laws about two major topics in physics: gravity and motion. His law of gravity deals with the attractive force (gravitation) that exists between any two objects that have mass. The strength of the force depends on how close they are (the nearer they are, the stronger is the force trying to pull them together) and how massive they are (the more massive they are, the stronger the force between them). In mathematical terms the gravitational force is proportional to the product of the masses and inversely proportional to the distance between them.

Newton's first law of motion states that an object at rest will remain at rest (and a moving object will continue moving) unless an outside force acts on it. According to the second law, force can be defined as something that makes an object accelerate (force equals mass times acceleration). The third law states that for every "action" (a force that one object exerts on another) there is an equal and opposite "reaction" (exerted by the second object on the first).

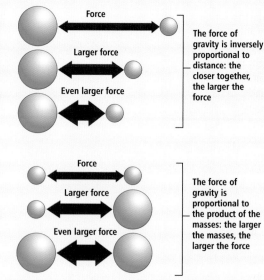

Force
Larger force
Even larger force

The force of gravity is inversely proportional to distance: the closer together, the larger the force

Force
Larger force
Even larger force

The force of gravity is proportional to the product of the masses: the larger the masses, the larger the force

Book I. Plate III. Part I.

In the branch of physics we now call optics, Newton's main studies concerned the nature of light. By allowing a narrow beam of white light from the Sun to pass through a glass prism, Newton split the light into a multicolored spectrum, the sequence of colors seen in a rainbow. He showed that white light is made up of a variety of colors. (Today we say that it is made up of many different wavelengths.) Telescopes of the day produced images that were surrounded by a spectrum of colors because the poor-quality lenses brought different colors into focus in different places. Newton overcame the problem by using mirrors instead of lenses and in 1668 produced one of the first reflecting telescopes with mirrors he made himself.

Newton was convinced that light is composed of a "flux" of minute particles ("corpuscles"). The theory was soon challenged by **Christiaan Huygens** (1629–95) and others, who postulated that light travels as waves. The argument raged until the 20th century, when physicists finally concluded that light

▲ If we are to believe the traditional story, Newton formulated his law of gravity when he saw an apple fall while sitting in an orchard.

▶ This diagram from Newton's book *Opticks* of 1704 shows a prism splitting a ray of white light into a spectrum.

has properties of both particles and waves; but this had to await the development of quantum theory.

In 1703 Newton was elected president of the Royal Society, and two years later he was knighted. As Sir Isaac Newton he continued to be showered with honors. His final tribute was a state funeral and burial in Westminster Abbey, London. His name lives on in the SI unit of force, which is called the newton (the force that gives 1 kilogram an acceleration of 1 meter per second per second).

CHEMISTRY AND PHYSICS

1694 Italian scientist Carlo Renaldini (1615–98) suggests that both the freezing point and boiling point of water should be used as "fixed points" on thermometers.

1697 German chemist **Georg Stahl** (1660–1734) champions the (incorrect) phlogiston theory of combustion devised by Johann Becher (1635–82) in 1669.

1700 German mathematician **Gottfried Leibniz** (1646–1716) founds the Berlin Academy, the first major national academy of science.

1701 Dutch chemist Wilhelm Homburg (1652–1715) discovers boric acid (sometimes called boracic acid).

1701 English astronomer **Edmond Halley** (1656–1742) publishes a map of the world showing magnetic variations.

Halley's world map of 1701, giving magnetic variations, proved to be of practical value to navigators and was used for many years.

BIOLOGY AND MEDICINE

c.1690 English naturalist John Ray (1627–1705) distinguishes between monocotyledons (plants with one seed leaf) and dicotyledons (plants with two seed leaves).

1691 English naturalist John Ray (1627–1705) suggests that fossils are the remains of creatures that lived many years ago in the distant past.

1691 English physician Clopton Havers (c.1655–1702) discovers Haversian canals, which are the fine channels carrying blood vessels and nerves that form a network within bone; he also publishes the first comprehensive book on bones of the body.

1694 German botanist Rudolf Camerarius (1665–1721) establishes the existence of male and female sexes in plants.

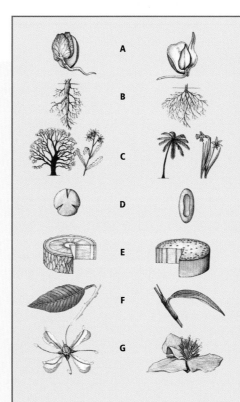

John Ray distinguished two main types of flowering plants, called monocotyledons and dicotyledons (often shortened to "monocots" and "dicots"). Examples of monocots are grasses such as rice, corn, wheat, and sugarcane. Dicots include roses, apple trees, and oak trees. A cotyledon is a seed leaf—monocots have one seed leaf, and dicots have two. The diagrams show other general distinguishing features of the two groups, with dicots on the left and monocots on the right.
Dicots: A: two seed leaves; B: often persistent taproot; C: spreading habit includes trees; D: pollen has three furrows, or pores; E: cylindrically arranged vascular tissues in stem often with secondary thickening (wood); F: leaf with branching veins and often with stalk; G: flower parts usually in fours or fives.
Monocots: A: one seed leaf; B: often fibrous root system; C: small herbs (exception palms); D: pollen has one furrow, or pore; E: vascular bundles scattered within stem and no true secondary thickening; F: leaf with parallel veins and sheathing at base; G: flower parts usually in threes or multiples of three.

ENGINEERING AND INVENTION

1690 French engineer **Denis Papin** (1647–1712) conceives the idea of a steam-powered paddleboat and constructs a primitive steam engine.

1692 In France the 46-mile (74-km) Orléans Canal from Orléans to Paris, connecting the Loire and the Seine rivers, is completed.

1692 Engineered by Pierre-Paul Riquet de Bonrepos (1604–80), the 32-mile (51.5-km) Canal du Midi in France is completed. It links the Mediterranean Sea with the Atlantic Ocean.

1694 English clockmaker Daniel Quare (1648–1724) makes a portable barometer.

1697 French engineer Paul Hoste (1652–1700) publishes *Théorie de la Construction des Vaisseaux* (Theory of Construction of Vessels).

1698 English engineer Henry Winstanley (1644–1703) completes the first lighthouse at Eddystone Rocks in the English Channel; in 1699 the height was increased because ocean spray was extinguishing the light.

1698 English mining engineer **Thomas Savery** (c.1650–1715) invents a steam pump, which becomes the forerunner of the atmospheric steam engine.

1700 Swedish engineer Christoph Polhem (1661–1751) improves metal rolling mills to produce bars with shaped profiles.

The first Eddystone lighthouse of 1698.

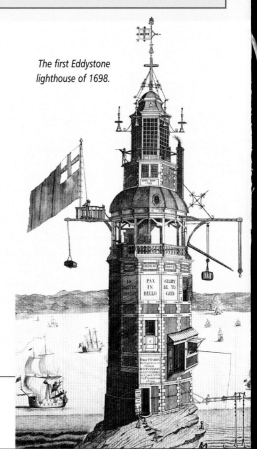

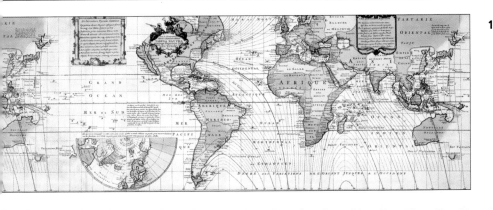

CHEMISTRY AND PHYSICS

1704 English scientist and mathematician *Isaac Newton* (1642–1727) publishes his book *Opticks*, in which he champions the corpuscular (particle) theory of light and explains the actions of lenses and prisms.

Rudolf Camerarius identified the reproductive parts of a plant and described their function in fertilization in 1694. The male organ is the stamen, and the female organs are the stigma, style, ovary, and ovule.

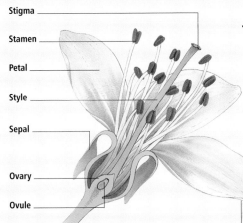

- Stigma
- Stamen
- Petal
- Style
- Sepal
- Ovary
- Ovule

c.1695 Danish physician Kaspar Bartholin the Younger (1655–1738) discovers Bartholin's glands, which are small mucus-producing glands in the reproductive system of female mammals.

1695 English botanist Nehemiah Grew (1641–1712) discovers the use of Epsom salts (magnesium sulfate) as a laxative.

1696 Dutch scientist **Antonie van Leeuwenhoek** (1632–1723) publishes a description of microorganisms (which he called "animalcules"); today we know most of them as protists.

1699 English botanist John Woodward (1665–1728) demonstrates that plants grow best if other substances are added to their water.

1701 Italian physician Giacomo Pylarini (1659–1715) inoculates three children in Constantinople with smallpox to prevent more serious disease when they are older; some people consider Pylarini to be the first immunologist.

1702 English anatomist William Cowper (1666–1709) discovers Cowper's glands, which are small mucus-producing glands in the male reproductive system.

BIOLOGY AND MEDICINE

1701 French mechanic Charles Plumier (1646–1704) publishes his book *L'Art de Tourner* (The Art of Turning), in which he describes a lathe for turning iron and shows the progress being made in metalworking machines.

1701 English agriculturist Jethro Tull (1674–1741) invents a mechanical seed drill for sowing seeds.

1703 The first Eddystone lighthouse is swept away in a violent storm. Winstanley achieved his wish of being in the lighthouse as it faced the strongest storm and died as a result.

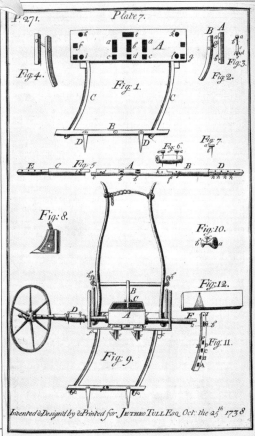

1703 English physicist Francis Hawksbee (c.1666–1713) invents an improved vacuum pump.

1704 Italian clockmaker Nicolas Fatio de Duiller (1664–1753) makes a clock with jewel bearings.

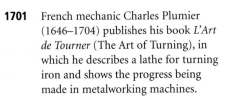
Before 1701 farmers had always sown seeds by hand. Jethro Tull invented a horse-drawn machine that could plant seeds in straight lines. It consisted of a wheeled vehicle that had a box filled with seed. A wheel-driven ratchet planted the seed out evenly as the drill was pulled across the field.

ENGINEERING AND INVENTION

1690–1704 A.D.

TELESCOPES

The first lenses were used mainly as magnifying glasses. They were convex (converging) lenses, curving outward on each side, that produced enlarged images of nearby objects. But scientists and astronomers needed to have an enlarged image of distant objects. The telescope fulfilled this need.

Hans Lippershey (*c*.1570–*c*.1619), the German-born Dutch eyeglass maker, made his first telescope in 1608. He sold his invention to the Dutch government for use by the military, but they would not grant him a patent because other people laid claim to the invention. News of the device—which would be described today as a refracting telescope—reached the Italian scientist **Galileo Galilei** (1564–1642), who immediately made his own telescopes to study the skies. Among his many discoveries were sunspots, craters on the Moon, and the four major moons of Jupiter.

Another contemporary astronomer, the German **Johannes Kepler** (1571–1630), explained correctly how this type of telescope works. A concave eyepiece lens focuses an enlarged image produced by the convex object lens. He also suggested that a telescope would provide a wider field of view using two convex lenses—a design successfully adopted by German astronomer Christopher Scheiner (1575–1650) in 1611. Called an astronomical telescope, it produces an upside-down image. For centuries afterward, images of the Moon, for instance, were always shown with "north" at the bottom.

Lenses of the time suffered from various defects, such as chromatic aberration, which results in colored fringes around images. Grinding and polishing lenses reduces aberration to some extent, as discovered by Dutch scientist **Christiaan Huygens** (1629–95) in 1655. With his improved astronomical telescope he first observed the rings of Saturn.

But it was not until 1758 that Englishman John Dollond (1706–61), maker of optical and astronomical instruments, finally made an achromatic telescope. He rediscovered a method of making achromatic lenses first used in 1733 by English amateur astronomer Chester Hall (1703–71). The method, still used today, involves making a compound lens with two separate components stuck together. The second component, made of crown glass, corrects the aberrations caused by the first component, which is made out of flint glass. It works because the two types of glass bend light rays in slightly different ways.

Another method of avoiding chromatic aberration involved using lenses with only slight curvature and therefore a long focal length (meaning the length of the telescope's light path from the main mirror or object lens to the focal point—the location of the eyepiece). This meant making telescopes very long. Telescopes measuring 33 feet (10 m) long were common, and by 1650 the Polish amateur astronomer Johannes Hevelius (1611–87) had constructed a telescope 148 feet (45 m) long. These so-called aerial telescopes employed only a skeleton tube hanging from a mast and were aimed using a system of ropes and pulleys.

A better way to view images is with a reflecting telescope that uses mirrors instead of lenses, because mirrors do not cause chromatic aberration. **James Gregory** (1638–75), a Scottish mathematician and inventor, realized this in 1663 when he published a design for a telescope that had a small, curved secondary mirror to reflect the light back through a hole in the primary mirror to an eyepiece.

▲ Isaac Newton's reflecting telescope. Scottish mathematician James Gregory designed the first reflecting telescope in 1663, and Isaac Newton built his own version in 1668. This telescope had the eyepiece at the side. It had a 1.3-inch (3.3-cm) mirror and magnified objects about 40 times.

KEY DATES

1608 First refracting telescope

1655 Huygens's improved refractor

1663 Gregory's reflecting telescope

1672 Cassegrain's improved reflector

1668 Newton's reflecting telescope

1758 Dollond's achromatic telescope

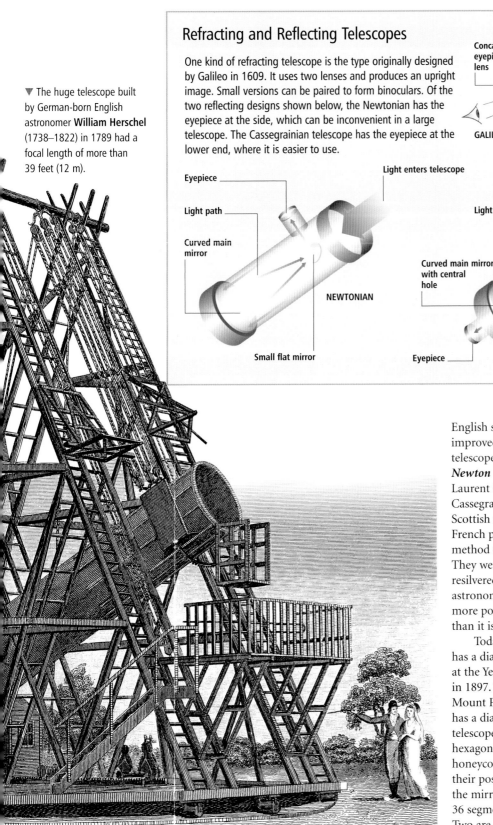

▼ The huge telescope built by German-born English astronomer **William Herschel** (1738–1822) in 1789 had a focal length of more than 39 feet (12 m).

Refracting and Reflecting Telescopes

One kind of refracting telescope is the type originally designed by Galileo in 1609. It uses two lenses and produces an upright image. Small versions can be paired to form binoculars. Of the two reflecting designs shown below, the Newtonian has the eyepiece at the side, which can be inconvenient in a large telescope. The Cassegrainian telescope has the eyepiece at the lower end, where it is easier to use.

Concave eyepiece lens
Final image
GALILEAN
Convex objective lens

Eyepiece
Light path
Curved main mirror
Light enters telescope
NEWTONIAN
Small flat mirror

Light enters telescope
Light path
Curved main mirror with central hole
CASSEGRAINIAN
Eyepiece
Small convex mirror

English scientist **Robert Hooke** (1635–1703) later improved the design. Other types of reflecting telescope were built—by English scientist *Isaac Newton* (1642–1727) in 1668 and by French priest Laurent Cassegrain (1629–93) in 1672. (The complex Cassegrainian design was not perfected until 1740 by Scottish optician James Short, 1710–68.) In 1857 French physicist Léon Foucault (1819–68) devised a method of silvering glass to make curved mirrors. They were easier to manufacture, and could be resilvered if accidentally damaged. Since then astronomical telescopes have been made larger and more powerful—it is far easier to make a big mirror than it is to make a big lens.

Today's largest refracting telescope, using lenses, has a diameter of about 40 inches (100 cm). It is sited at the Yerkes Observatory near Chicago and was built in 1897. The mirror in the Hale reflecting telescope at Mount Palomar, California, commissioned in 1948, has a diameter of 16.6 feet (5 m). Even larger telescopes have mirrors made up of many smaller hexagonal segments that fit together like a honeycomb. A computer controls the segments, and their positions can be altered to adjust the focus of the mirror. The largest telescopes of this type, with 36 segments each, have mirrors 33 feet (10 m) across. Two are situated at the Keck Observatory in Hawaii.

ASTRONOMY AND MATH

1706 Welsh mathematician William Jones (1675–1749) introduces the symbol π (Greek letter pi) for the ratio of the circumference of a circle to its diameter (π = approximately 3.1416).

1712 English mathematician Brook Taylor (1685–1731) proposes Taylor's theorem, a fundamental method that can be used to find the sum of a mathematical series.

1712 First volume of *Historia Coelestis Britannica* (cataloguing the position of 3,000 stars) by English astronomer John Flamsteed (1646–1719) is published without his permission.

1712 Italian mathematician Giovanni Ceva (c.1647–c.1734) publishes *De Re Numeraria* (Concerning Money Matters), the first clear application of math to economics.

1717 English astronomer Abraham Sharp (1651–1742) calculates the value of π (pi) to 72 decimal places.

1718 French astronomer Jacques Cassini (1677–1756), in a joint publication with his father **Giovanni Cassini** (1625–1712), confirms Descartes's (incorrect) prediction that the Earth is elongated at the poles.

CHEMISTRY AND PHYSICS

1709 English physicist Francis Hawksbee (c.1666–1713) describes capillary action, the phenomenon that causes a liquid to rise up a very narrow tube and a sponge or blotting paper to soak up liquids.

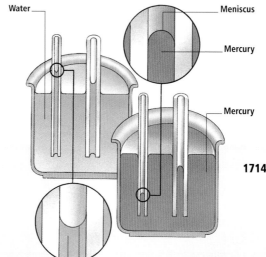

Water — Meniscus — Mercury — Mercury — Water — Meniscus

Capillary action was first described in 1709. The curved surface of a liquid in a capillary tube is called the meniscus. In a less dense liquid such as water, the liquid rises up the tube and forms a concave meniscus (it curves inward). In the case of a dense liquid such as mercury, capillary action forces the liquid downward, and the meniscus is convex (curved outward).

BIOLOGY AND MEDICINE

1707 English physician John Floyer (1649–1734) produces a special watch for counting patients' pulse rates.

1711 Italian naturalist Luigi Marsigli (1658–1730) shows the animal nature of corals (formerly held to be plants).

1714 French physician Dominique Anel (1679–1730) invents a fine-point syringe for use in treating fistula lacrymalis (an opening in the cheekbone near the eye).

ENGINEERING AND INVENTION

1706 English physicist Francis Hawksbee (c.1666–1713) constructs an electrostatic generator.

1708 German alchemist Johann Böttger (1682–1719) invents hard-paste porcelain (previously the secret of making porcelain was known only to Chinese craftsmen).

1709 English iron founder **Abraham Darby** (c.1678–1717) introduces the use of coke for iron smelting (previously only expensive charcoal could be used).

1709 Polish-born Dutch physicist Gabriel Fahrenheit (1686–1736) invents the alcohol thermometer and the Fahrenheit temperature scale.

1709 Brazilian Jesuit priest Bartholomeu de Gusmao (1685–1724) makes a model hot-air balloon.

1710 German printer Jakob Le Blon (1667–1741) develops a three-color mezzotint printing process.

1712 English engineer **Thomas Newcomen** (1663–1729) invents an atmospheric steam engine that employs a piston, unlike the engine built by **Thomas Savery** (c.1650–1715) in 1698.

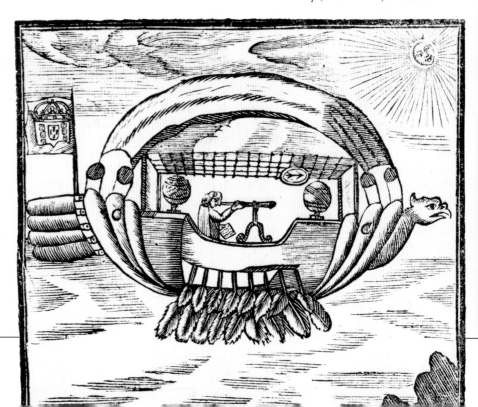

A fanciful drawing of Gusmao's hot-air balloon.

See also The Development of Printing **3**:28–29; Galileo Galilei **4**:8–9; Iron Smelting **4**:32–33; Navigation at Sea **4**:36–37

1718 English astronomer **Edmond Halley** (1656–1742) identifies stellar proper motion, which is the apparent movement of a star on the celestial sphere. It results from the star's very gradual movement relative to the Sun (not relative to the Earth). Stars with the largest proper motions are closest to the Earth.

1710 French chemist René-Antoine Ferchault de Réaumur (1683–1757) presents to the Academy of Sciences in Paris a material completely woven from glass fiber.

1717 Italian physician Giovanni Lancisi (1654–1720) blames malaria on the bite of the mosquito in his book *De Noxiis Paludum Effluviis* (Concerning the Noxious Effluvia of Marches).

1718 French mathematician Abraham de Moivre (1667–1754) produces *The Doctrine of Chances*, his first book on probability.

1719 English mathematician Brook Taylor (1685–1731) publishes *New Principles of Linear Perspective*, in which he demonstrates the principle of the vanishing point.

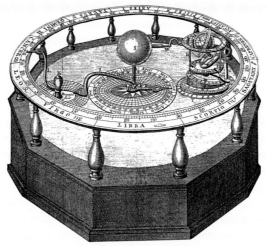

Perspective is based on the idea of the vanishing point (VP), a point on the horizon toward which all parallel lines appear to lead. The horizon is always on the viewer's eye level. Perspective provides a way of representing two- or three-dimensional objects on a two-dimensional surface (such as a sheet of paper). A tiled floor, for example, can be drawn in one-point perspective, while cubes have to be represented in two- or three-point perspective.

VP

Three-point perspective

VP

One-point perspective

VP

VP

VP

VP

Two-point perspective

1712 English inventor George Graham (1673–1751) constructs a clockwork orrery, which is a working clockwork model of the solar system. Three years later he invents the deadbeat escapement for clocks.

1714 Polish-born Dutch physicist Gabriel Fahrenheit (1686–1736) makes a mercury thermometer.

1714 English engineer Henry Mill patents the first typewriter, but today little is known of its method of working.

1714 The British government offers a prize of £20,000 to the person who devises a method of finding longitude accurately at sea. In 1759 English clockmaker John Harrison (1693–1776) claims the prize but has to wait 14 years to be paid in full.

1715 English clockmaker John Harrison (1693–1776) invents a clock that runs for eight days on a single winding.

1715 French engineer Hubert Gautier (1660–1737) brings out a second edition of *Traité de la Construction des Chemins* (Treatise on Road Building).

1716 English astronomer **Edmond Halley** (1656–1742) invents the diving bell, so that workmen can build foundations underwater.

An ornate 18th-century clockwork orrery.

1716 A lighthouse in Boston Harbor is the first to be constructed in North America.

1716 J. N. de la Hire invents a double-acting water pump, which produces a continuous stream of water.

1716 French engineer Hubert Gautier (1660–1737) publishes his book *Traité de la Construction des Ponts* (Treatise on Bridges), which has a lasting influence on bridge design.

1718 English inventor James Puckle patents a flintlock semi-automatic cannon.

1719 German printer Jakob Le Blon (1667–1741) develops a four-color printing process based on superimposing the primary colors blue, yellow, and red plus black.

1705–1719 A.D.

THE NATURE OF LIGHT

*E*arly scientific studies revealed various properties of light—how it is bent by lenses, how it casts shadows, even how fast it travels—but fundamental to understanding those properties is a knowledge of the nature of light itself. In particular, does light consist of a stream of minute particles, like bullets from a machine gun? Or does it consist of waves capable of rippling across the vast vacuum of space?

▲ Isaac Newton was one of the first people to make a scientific study of light. He believed that light travels as a stream of particles.

It is clear that parallel rays of light bend as they go through a lens and come to a focus. The concentration of the rays allows a magnifying glass to be used as a burning glass, an application known since ancient times. In 212 B.C. Greek scientist *Archimedes* (*c.*287–212) is said to have used a burning glass to destroy ships of the Roman fleet at Syracuse. The first person to measure the bending of light in this way was the Dutch mathematician Willebrord Snell (1580–1626). In 1621 he found that when a ray of light goes through a piece of glass, the angle of incidence (at which it enters the glass) is related to the angle of refraction (the angle through which it is bent) by a property of the glass now known as the refractive index.

It was another mathematician, the Frenchman Pierre de Fermat (1601–65), who figured out how light casts shadows. He stated that it is because light always travels in straight lines—it will not "go around corners" to illuminate a shadow. Known as Fermat's principle, it was proposed in 1640. He also observed that light travels slower in a denser medium.

The first attempt to measure the speed of light was made in 1676 by Danish astronomer Ole Rømer (1644–1710). He was checking the predictions made by Italian astronomer **Giovanni Cassini** (1625–1712) about the timing of eclipses of Jupiter's moons (when they move out of sight behind the planet). He discovered that the eclipses seemed to happen earlier than predicted when the Earth was moving toward Jupiter and later when the Earth was moving away from it. Rømer accounted for the differences by assuming that the light had to travel a shorter or longer distance, and that light must therefore have a finite speed, which he calculated as 140,000 miles (225,000 km) per second—about 75 percent of the actual value. It was nearly 200 years before French physicist Armand Fizeau (1819–96) obtained a more accurate value—about 5 percent too high—of 195,737 miles (315,000 km) per second. This in turn was improved on by American physicist Albert Michelson (1852–1931) in 1882, who calculated it to be 186,325 miles (299,853 km) per second. The value that is used today is 186,288 miles (299,793 km) per second.

Measuring the Speed of Light

In 1849 Armand Fizeau made the first fairly accurate measurement of the speed of light. He bounced a light ray between two mirrors 5.6 miles (9 km) apart, directing the ray between the teeth of a fast-rotating cogwheel. The returning ray went between the next pair of teeth and then through a semisilvered mirror to the observer. He adjusted the speed of the cogwheel so that there was no flicker as the light traveled the 5.6 miles (9 km) and back. Using the wheel's speed and the distance traveled by the light, Fizeau was able to calculate the speed of light.

Albert Michelson's method of 1882 used a rotating mirror to reflect a light beam to a curved mirror 21.7 miles (35 km) away. The first mirror was rotated by an electric motor and reflected the returning beam into an eyepiece. The speed of the motor was gradually increased until the light did not flicker. The time taken for the light to make the 44-mile (70-km) round trip could then be calculated using the rotation speed of the mirror.

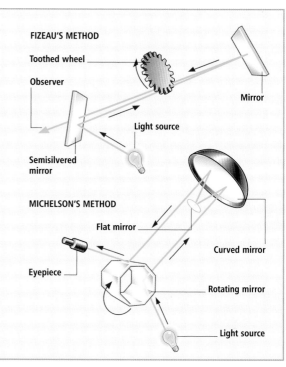

FIZEAU'S METHOD

Toothed wheel

Observer

Mirror

Light source

Semisilvered mirror

MICHELSON'S METHOD

Flat mirror

Curved mirror

Eyepiece

Rotating mirror

Light source

▲ In about 212 B.C. Greek scientist Archimedes is said to have used a lens as a burning glass to defend Syracuse against attacking Roman ships.

Snell's Law

Snell's law is named for the Dutch mathematician Willebrord Snell (1580–1626), who discovered it in 1621 when he was professor of mathematics at Leyden (now Leiden) University. It concerns refraction of light, which is the way a ray of light changes direction when it goes from one transparent medium to another—for example, moving from air into a block of glass. The amount of refraction depends on an optical property of the denser medium, called its refractive index. The law says that the sine of the angle of incidence divided by the sine of the angle of refraction equals the refractive index.

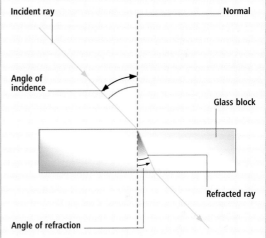

In 1675 English scientist *Isaac Newton* (1642–1727) postulated that light travels as a stream of minute particles ("corpuscles"). Over the years various physicists challenged this idea, the first being his great rival **Robert Hooke** (1635–1703), who had already proposed the wave theory in 1665. The fact that light is refracted by glass and the fact that it travels more slowly in water than it does in air, for example, were cited as evidence that it travels as waves. The final nail in the coffin of the corpuscular theory came in 1801, when English physicist Thomas Young (1773–1829) discovered the interference of light, a phenomenon in which white light shining through a narrow slit is split into the colors of the rainbow. At that time it could be explained only if light was assumed to travel as waves. Young published his findings in 1804.

That is how the argument remained until the beginning of the 20th century, when German physicist **Max Planck** (1858–1947) put forward his quantum theory. It postulates that all forms of energy, including light, travel in finite "packets," or quanta, similar in fact to Newton's corpuscles. But as modern physics continued to develop, French physicist **Louis de Broglie** (1892–1987) suggested in 1924 that moving particles can also behave like waves, and this was soon proved to be the case. So Newton, Hooke, and the others were all correct, and one of the great arguments of science melted away.

▼ Interference is a phenomenon that demonstrates the wave properties of light. The diagram below shows how white light shining through a pair of narrow slits is split into its component colors, which combine to produce a pattern of colored fringes, as can be seen in a soap bubble or in an oil film on water.

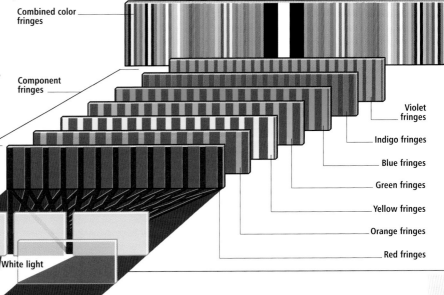

Combined color fringes

Component fringes

Violet fringes

Indigo fringes

Blue fringes

Green fringes

Yellow fringes

Orange fringes

Red fringes

White light

ASTRONOMY AND MATH

1727 Swiss mathematician Leonhard Euler (1707–83) introduces the symbol e (for the base of natural logarithms). He also discovers various relationships involving e.

1729 English astronomer James Bradley (1693–1762) discovers the aberration of starlight, a small apparent shift in a star's position that results from a combination of the motion of the Earth and the fact that light travels at a finite speed.

1731 English astronomer John Bevis (1695–1771) discovers the Crab Nebula. Later (1758) it is rediscovered by French astronomer Charles Messier (1730–1817), who gives it the designation M1 and lists it first in his Messier Catalog. Today all nebulas have Messier numbers.

CHEMISTRY AND PHYSICS

1724 Dutch scientist Hermann Boerhaave (1668–1738) publishes his book *Elementa Chemiae* (Elements of Chemistry), the first major chemistry textbook.

1724 The St. Petersburg Academy of Sciences is founded in Russia by Peter the Great (1672–1725).

1725 German physician Johann Schulze (1684–1744) observes that daylight turns certain silver salts dark (later to have significance in photography).

1729 English physicist Stephen Gray (1666–1736) distinguishes between electrical insulators and conductors.

The house near Philadelphia belonging to John Bartram, the "Father of American botany."

BIOLOGY AND MEDICINE

1721 American physician Zabdiel Boylston (1676–1766) carries out the first smallpox inoculation in the U. S.

1727 English botanist Stephen Hales (1677–1761) writes *Vegetable Staticks*, the first book on plant physiology.

1728 The first botanical garden in the U. S. is opened by naturalist and explorer John Bartram (1699–1777) at his home near Philadelphia.

ENGINEERING AND INVENTION

The pianoforte (meaning "soft-loud") was a development of the harpsichord. Both instruments were based on a harp lying on its side. In a harpsichord the string is plucked by a small plectrum (originally made of quill), but the instrument could not make a contrast between loud and soft notes. However, the strings of a piano are struck by a felt-covered hammer, producing a very different sound and giving performers control over both the volume and length of a given note.

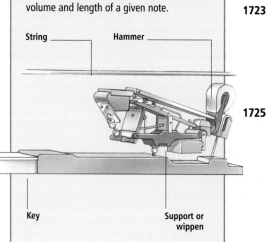

String   Hammer

Key   Support or wippen

c.1720 English clockmaker Christopher Pinchbeck (*c.*1670–1732) produces the copper–zinc alloy pinchbeck. It resembles gold, and soon after it is used widely in making watches and jewelry.

1720 Italian harpsichord maker Bartolomeo Cristofori (1655–1731) invents the pianoforte.

1723 French engineer Nicolas Bion (1653–1733) produces a description of the various surveying instruments in use at the time. Many of them remain unchanged for more than 200 years.

1725 French clockmaker Antoine Thiout (b.1692) constructs a clock that displays solar time. He calls it the equation clock (the equation of time is the difference between solar time, as shown by a sundial, and time based on the Earth's rotation on its axis, as shown by an ordinary clock).

1725 Scottish goldsmith William Ged (1690–1749) introduces stereotype printing, in which a mold is cast from a complete page of type and used to make a printing plate.

1726 English inventor George Graham (1673–1751) invents the mercury pendulum for clocks, which does not change in length with changes in temperature.

1728 English clockmaker John Harrison (1693–1776) creates the gridiron pendulum for clocks. Like George Graham's invention, its length is not affected by variations in temperature.

1728 French dentist Pierre Fauchard (1678–1761) invents the first dental drill and introduces fillings.

1729 Spanish-born French engineer Bernard Forest de Belidor (1698–1761) publishes a book of engineering tables and construction principles, *Science of Engineers*.

1733 French mathematician Abraham de Moivre (1667–1754) discovers the normal (bell-shaped) distribution curve, soon to become of major importance in statistical studies.

The Crab Nebula, discovered in 1731, is the remains of a supernova that exploded in 1054.

1730 French scientist Pierre Bouguer (1698–1758) demonstrates the Bouguer anomaly: that gravitational attraction decreases with altitude.

1733 French chemist Charles du Fay (1698–1739) distinguishes two types of static electricity: "vitreous" (positive) and "resinous" (negative).

1730 French surgeon George Martine (1702–41) performs the first tracheostomy, an operation to make an opening in the trachea (windpipe) to help a patient with life-threatening breathing difficulties.

1731 English agriculturist Jethro Tull (1674–1741) publishes *The New Horse-Hoeing Husbandry*, a book that recommends harrowing the soil, growing crops in rows, removing weeds by hoeing, and using manure as fertilizer.

1734 French chemist René-Antoine Ferchault de Réaumur (1683–1757) founds the science of entomology with his book *Mémoires pour Servir à l'Histoire des Insectes* (Memoirs Serving as a Natural History of Insects).

English clockmaker John Harrison.

1730 English mathematician John Hadley (1682–1744) devises the quadrant, a navigational instrument that was the forerunner of the sextant. A year later (1731) he produces an improved instrument called the octant.

1730 French chemist René-Antoine Ferchault de Réaumur (1683–1757) makes an alcohol thermometer and devises the Réaumur temperature scale (with zero set at the freezing point of water and its 80° mark set at the boiling point of water).

1732 French physicist Henri Pitot (1695–1771) creates the Pitot tube, an instrument for measuring speed of air flow. In the 20th century it is used as an air-speed indicator for aircraft.

1733 English engineer John Kay (1704–64) patents the flying shuttle, which greatly increases the speed at which a loom makes cloth.

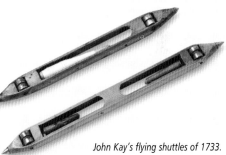

John Kay's flying shuttles of 1733.

1733 English amateur astronomer Chester Hall (1703–71) develops a simple achromatic lens for telescopes. (An achromatic lens does not produce colored fringes around the image.)

1733 The world's first technical college opens in Schemnitz, Hungary, as a mining academy.

1734 Swedish scientist and philosopher Emanuel Swedenborg (1688–1772) publishes *Regnum Subterraneum* (Mineral Kingdom), two volumes that describe techniques for mining and smelting metals.

ASTRONOMY AND MATH

CHEMISTRY AND PHYSICS

BIOLOGY AND MEDICINE

ENGINEERING AND INVENTION

1720–1734 A.D.

31

IRON SMELTING

▲ The bridge over the Severn River at Coalbrookdale was built by Abraham Darby III in 1779. It was the first ever cast-iron bridge, made of prefabricated cast-iron sections. It was a commercial venture, and travelers paid a toll to cross the bridge.

KEY DATES
1709 Coke in blast furnaces
1779 Iron bridge at Coalbrookdale
1828 Neilson's hot-air process
1857 Hot-blast stove

Iron has been known since ancient times, and the period from about 1100 B.C. is known as the Iron Age in the Middle East and Europe. But iron tools and weapons were rare and reserved only for rich people until methods of smelting iron from its ores became widespread after the invention of the blast furnace in about 700 A.D.

The ancient Egyptians got their iron from meteorites picked up in the desert, but by about 1350 B.C. they had developed methods of welding it in a hot fire. At about the same time, the Hittites in Anatolia (modern Turkey) began using iron, and knowledge of ironworking spread to India and China. Ancient Greeks used iron bolts to join blocks of stone, and in about 400 B.C. Chinese craftsmen made statues from a type of cast iron that had a low melting point.

The first blast furnace for iron, the Catalan forge in Spain, is thought to date from about 700 A.D. Another was constructed about a century later in Scandinavia. The principle involves simple chemistry, although, of course, that was unknown to the people of the time. The diagram below shows how it worked. The metalworkers made a furnace by digging a hole in the ground and lining it with packed earth and charred reeds—a form of fine charcoal. They added a conical chimney made of clay and slag (impurities from a previous furnace). The furnace was filled, or "charged," with a mixture of iron ore, limestone, and

charcoal, and set on fire. When it was hot, workers pumped bellows to force a blast of air through the furnace. The ore (iron oxide) changed into metallic iron by the action of carbon monoxide (CO) formed in turn by the action of the air on the charcoal (carbon). The main role of the limestone (calcium carbonate) was to form a slag with the silicate rock impurities in the ore. The slag floated on top of the molten iron, which could be tapped through a hole near the bottom of the furnace.

By the 14th century England was Europe's main iron-producing country. Waterwheels powered the bellows to produce a continuous stream of air for the blast furnaces, which could produce up to 3.3 tons (3 tonnes) of iron a day. This output required large amounts of charcoal, produced by burning wood. As a result, most of Britain's forests were destroyed. Then in 1709 English iron founder **Abraham Darby** (*c*.1678–1717) began using coke (derived from coal) instead of charcoal. Darby's compatriot Dud Dudley (1599–1684) claimed to have previously used coal in a blast furnace, but that is unlikely because the sulfur present in the coal would have spoiled the iron. Darby's development had a dramatic effect on the production and uses of cast iron, and cast-iron pans, pots, and kettles soon became commonplace in every home in England.

Darby built his furnaces at Coalbrookdale on the banks of the Severn River. In 1742 his son Abraham Darby II (1711–63) installed a steam engine to pump

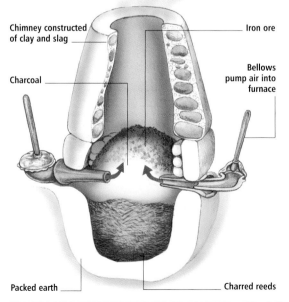

Chimney constructed of clay and slag

Charcoal

Iron ore

Bellows pump air into furnace

Packed earth

Charred reeds

▶ Early metalworkers used simple blast furnaces to smelt iron. First they dug a hole in the ground and built a chimney over it. They placed iron ore on top of smoldering charcoal in the pit. Using hand bellows, they forced air into the furnace, which raised the temperature and produced molten iron.

▲ The blast furnaces at Coalbrookdale ran nonstop for 24 hours a day. This painting, dating from 1801, shows the fiery glow in the sky as iron is drawn off from one of the furnaces.

water from the river to power the bellows. His grandson Abraham Darby III (1750–91) took over the company in 1768 and built a lasting memorial to the family. In 1779, using prefabricated cast-iron sections, he completed a bridge over the Severn River at Coalbrookdale. It is 98 feet (30 m) long and stands 39 feet (12 m) above the water. It still carries foot passengers, but it was closed to traffic in 1934.

The final improvements to the blast furnace came in the 1800s. In Glasgow, Scotland, in 1828 Scottish engineer James Neilson (1792–1865) improved its efficiency by preheating the air by sending it through a red-hot tube. The tube was heated at first by a coal fire and later by coal gas—a byproduct from coking furnaces. English inventor Edward Cowper (1819–93) improved Neilson's design in 1857 with his hot-blast stove, which used waste gases from the blast furnace itself to preheat the air.

◀ A selection of furnaces and their components form a border around this general view of a rolling mill. The mill rolled down bars of red-hot iron for making rails and thinner bars.

1735–1749 A.D.

ASTRONOMY AND MATH

1735 Swiss mathematician Leonhard Euler (1707–83) solves the Königsberg Bridges problem.

1735 English scientist George Hadley (1685–1768) proposes that the overall circulation of air in the atmosphere is due to convection currents.

1736 French mathematician Pierre de Maupertuis (1698–1759) states correctly that the shape of the Earth is an oblate spheroid (a sphere flattened slightly at the poles).

1736 French surveyor Alexis Clairaut (1713–65) measures the length of 1 degree of meridian (longitude), thus enabling accurate calculation of the size of the Earth.

1743 French astronomer Philippe de Chéseaux (1718–51) and Dutch amateur astronomer Dirk Klinkenberg (1709–99) independently discover the comet known as Comet de Chéseaux.

CHEMISTRY AND PHYSICS

1735 Spanish scientist and naval officer Antonio de Ulloa (1716–95) rediscovers platinum in South America. It had been used by local people for centuries and was found in Central America by Spanish explorers, but its discovery had not been announced.

1735 English physicist Stephen Gray (1666–1736) postulates that lightning is an electrical phenomenon.

1737 Swedish chemist Georg Brandt (1694–1768) discovers cobalt, the first discovery of a completely new metal since ancient times.

BIOLOGY AND MEDICINE

1735 Swedish naturalist **Carolus Linnaeus** (1707–78) publishes *Systema Naturae*, in which he classifies objects into three kingdoms: animal, plant, or mineral. Later (1749) he introduces binomial naming (genus and species names).

Linnaeus's Systema Naturae of 1735 classified plants according to the structure of their flower parts.

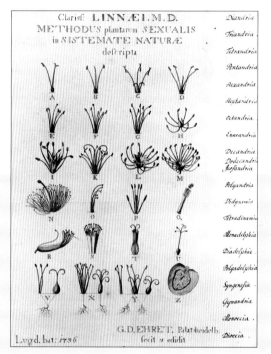

1736 French explorer Charles-Marie de la Condamine (1701–74) discovers India rubber (then called *caoutchouc*). Four years later he discovers the powerful paralyzing drug curare (used as an arrow poison by South American Indians).

ENGINEERING AND INVENTION

1735 English clockmaker John Harrison (1693–1776) unveils his first chronometer, a clock that keeps time well enough to be used to calculate longitude at sea.

1738 English inventor Lewis Paul (d.1759) produces a machine for carding wool. (Carding is the process of combing out woolen fleeces into parallel fibers.) With English engineer John Wyatt (1700–66) Paul went on to invent a water-powered spinning machine.

1738 English metallurgist William Champion (1710–89) devises a new industrial process for the large-scale production of zinc from its ores.

1740 English metallurgist Benjamin Huntsman (1704–76) invents the crucible process for making steel in batches.

1741 Swedish engineer Christoph Polhem (1661–1751) introduces a gear-cutting machine.

1742 American scientist and politician *Benjamin Franklin* (1706–90) invents a wood-burning stove. The design is patented, and later stoves of the same design come to be known as "Franklin stoves."

1742 French metallurgist Jean-Paul Malouin (1701–78) creates a process for covering steel with a layer of zinc (later to be called galvanizing).

1743 English metalworker Thomas Boulsover (1706–88) produces "Sheffield plate," metalware consisting of copper coated with a thin layer of silver.

The city of Königsberg (once part of Germany but now in Russia and known as Kaliningrad) is built on a river that divides around an island. Seven bridges span the river in various places. The problem is to find a continuous route that crosses all seven bridges but only once each. Leonhard Euler showed mathematically in 1735 that it could not be done. This is a problem in topology, the branch of math that deals with the general properties of space and shapes. In topology twisting or stretching (called continuous deformations) have no effect on the shape's properties. A circle and an ellipse, for example, are topologically equivalent because they can be changed into each other by continuous deformation.

ASTRONOMY AND MATH

1743 French mathematician Jean d'Alembert (1717–83) establishes mathematical dynamics (a branch of mechanics) with his book *Traité de Dynamique* (Treatise on Dynamics). Three years later he advances the theory of complex numbers. (Complex numbers have both a "real" and an "imaginary" part—known as imaginary in mathematics because they involve *i*, the "imaginary" square root of –1.) A year later (1747) Alembert introduces partial differential equations.

1743 English mathematician Thomas Simpson (1710–61) devises Simpson's rules, a systematic approach to finding the area bounded by a curve.

1748 English astronomer James Bradley (1693–1762) discovers the nutation of the Earth, which is the slight "nodding" of the Earth's axis as it describes a very slow circle in space.

1738 Swiss scientist Daniel Bernoulli (1700–82) becomes the first to put forward a kinetic theory of gases—that their properties can be explained by considering gases to be composed of rapidly moving small particles of matter.

1742 Swedish astronomer Anders Celsius (1701–44) devises the 100-degree Celsius temperature scale (later known as the centigrade scale but now called by its original name).

1743 The American Philosophical Society is founded.

1745 Russian scientist Mikhail Lomonosov (1711–65) compiles a catalog of more than 300 minerals.

1747 German chemist Andreas Marggraf (1709–82) discovers sugar in beets.

CHEMISTRY AND PHYSICS

1740 Swiss naturalist Charles Bonnet (1720–93) observes parthenogenesis in aphids (parthenogenesis is the phenomenon in which unfertilized female animals give birth to young).

1745 French surgeon Jacques Daviel (1693–1762) successfully performs an operation for the removal of a cataract from a patient's eye.

1747 Scottish physician James Lind (1716–94) experiments with citrus fruits to prevent scurvy among sailors in the British Royal Navy (citrus fruits are rich in vitamin C, the lack of which is now known to cause scurvy).

1748 Scottish physician John Fothergill (1712–80) gives the first description of diphtheria.

BIOLOGY AND MEDICINE

Brass rod joined to chain, touching the inner foil

Rubber stopper

Inner foil

Glass jar

Outer foil

Chain

The Leyden jar is named for the Dutch University of Leyden (now Leiden) where it was invented in 1745 by Pieter van Musschenbroek. It is a type of condenser used for storing electrical charge. It has plates of metal foil glued inside and outside a glass jar. A brass rod running through an insulating stopper has at its end a short length of metal chain that makes contact with the inner foil. An electrostatic charge, produced by rubbing an insulating rod or by an electrostatic generator, can be brought up to the brass rod to charge up the condenser. A person who touches the brass knob receives a powerful electric shock. According to one account, French cleric and physicist Jean-Antoine Nollet connected a charged Leyden jar to a row of monks holding hands and watched them all leap into the air with the shock.

1745 Dutch physicist Pieter van Musschenbroek (1692–1761) invents the Leyden jar, a simple form of electrical condenser.

1746 English chemist John Roebuck (1718–94) develops the lead-chamber process for making sulfuric acid.

1747 French monk and physicist Jean-Antoine Nollet (1700–70) devises an electrometer (an instrument for measuring electrical charge).

ENGINEERING AND INVENTION

NAVIGATION AT SEA

The crew of a ship in the middle of an open ocean needs to know what direction it is going in and exactly where it is. A compass can indicate direction, and the magnetic compass was in regular use by the 1100s. But accurate positioning needs a knowledge of latitude and longitude, which proved to be much more difficult to determine.

▲ John Harrison's fifth chronometer gained only 4.5 seconds in ten weeks. It was more accurate even at sea than any other clock on land.

Latitude indicates a position in terms of its distance north or south of the equator. It is measured in degrees. For example, Philadelphia is at a latitude of 40° north. Latitude can be found by measuring the angle of a particular heavenly body above the horizon and consulting books of tables or almanacs. At night the angle of the polestar or during the day the angle of the Sun at noon can be measured and compared with tables. Early sailors had various instruments for measuring these angles. Using a cross-staff, a sailor sighted along a 3-foot- (1-m-) long staff while moving a crosspiece until the lower end lined up with the horizon and the upper end

coincided with the star or the Sun. The staff was calibrated in degrees from which the sailor could read off the angle. It was first described in about 1330 by French astronomer Levi ben Gershom (1288–1344) and used in Europe until the 18th century.

In 1594 English sailor John Davis (c.1550–1605) invented the backstaff. It was pointed in the opposite direction, and the operator did not need to look directly into the Sun. The quadrant was a similar instrument, used also by astronomers and by gunners to set the correct angles for aiming artillery pieces.

Then in 1731 English mathematician John Hadley (1682–1744) invented the octant, incorrectly named Hadley's quadrant at the time. Anglo-American inventor Thomas Godfrey (1704–49) of

KEY DATES

1594 Backstaff

1731 Octant

1735 Chronometer

1757 Sextant

1759 Harrison's prize-winning chronometer

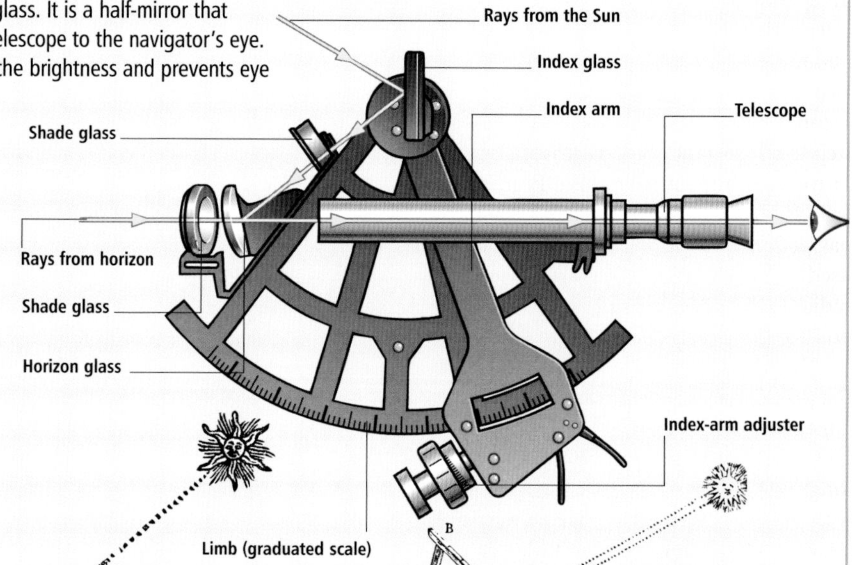

▼ Early instruments for measuring the angle of the Sun in the sky, essential for navigation, included (left to right) the cross-staff, the back-staff, and the quadrant.

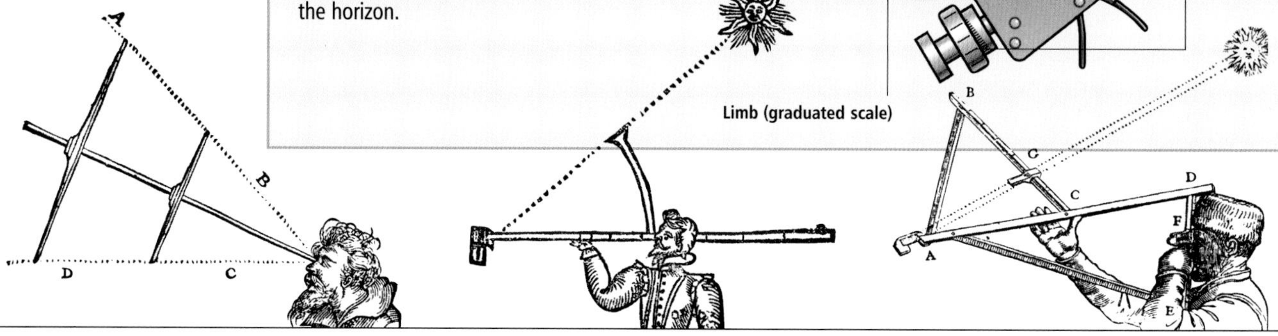

The Sextant

A navigator uses a sextant to measure the angle of the Sun (or at night a prominent star) above the horizon. Special tables convert the angle into the navigator's latitude, the angular distance north or south of the equator. The index glass (in fact, a mirror) reflects the Sun's rays onto a second mirror called the horizon glass. It is a half-mirror that reflects the rays along a telescope to the navigator's eye. A shade glass cuts down the brightness and prevents eye damage. The navigator also looks through the plain (unsilvered) half of the horizon glass at the horizon and adjusts the angle of the index glass until the Sun's image appears to be on the horizon. The graduated scale on the limb of the sextant then indicates the angle of the Sun above the horizon.

Rays from the Sun

Index glass

Index arm

Telescope

Shade glass

Rays from horizon

Shade glass

Horizon glass

Index-arm adjuster

Limb (graduated scale)

They were published in 1766 in the *Nautical Almanac* by English astronomer Nevil Maskelyne (1732–1811) and were subsequently revised every year.

The solution to the longitude problem lay in finding an accurate way of measuring time, which varies locally depending on longitude. For example, at 12 noon in London, England, it is 7.00 a.m. in Philadelphia (longitude about 75° west). So if we know the exact time at a given place when it is noon in London, we can calculate its longitude. To do this we need a chronometer, a very accurate clock. In 1714 the British government offered a prize of £20,000 to anyone who could produce such an instrument. A condition of the competition was that the "sea clock" had to gain or lose no more than 2 minutes after a six-week voyage to the West Indies and back.

English clockmaker John Harrison (1693–1776) took up the challenge and in 1735 introduced his first chronometer. But it was his fourth instrument of 1759 that won the prize (or half of it, since the government retained half the money until Harrison could prove that the chronometer could be duplicated). He did not receive the remainder until 1773, and then only after King George III pleaded Harrison's case.

◀ Finding latitude in the 1600s, before the invention of the sextant, involved taking a sighting on the Sun or a star. The task was made even more difficult on the pitching deck of a ship.

Philadelphia invented an almost identical instrument independently. In the octant a pivoted arm carries a mirror that can be moved to bring an image of the Sun in line with another mirror. The second mirror also gives a view of the horizon. The maximum angle it could measure was 45°. From there it was a simple step to the sextant (which measured up to 60°), introduced by Scottish naval officer John Campbell (*c.*1720–90) in 1757. It remained the standard navigational instrument for 250 years. It was even used on aircraft until it was finally supplanted by radio beacons and the satellite-based GPS (global positioning system).

In 1884 an international conference agreed that the prime meridian (longitude 0°) should be the Greenwich meridian that runs through Greenwich Observatory in London. However, finding the longitude of any other place—its position east or west of the Greenwich meridian (longitude 0°)— proved to be far more difficult than calculating latitude. For centuries sailors measured the angle between the Moon and another heavenly body, and consulted tables called ephemerides that gave the day-to-day positions of the Moon. German astronomer Johann Müller (1436–76), also known as **Regiomontanus**, drew up the first tables in 1474.

Finding Longitude

These diagrams show how to use a chronometer (a very accurate clock) to find a ship's longitude—its position east or west of the prime meridian at longitude 0°. The ship sails from Greenwich on the prime meridian at 12 noon local time. The chronometer is also set to 12 o'clock. After sailing west for five days, at 12 noon local time (easily gauged by using a sextant to plot the Sun's highest point) the chronometer reads 4 o'clock in the afternoon. In other words, local time is four hours behind Greenwich time, as shown by the chronometer. The Earth has rotated on its axis for four hours since it was 12 noon in Greenwich. The four hours represent $^4/_{24}$, or $^1/_6$, of a complete rotation, that is, $^1/_6$ of 360,° or 60°. This ship's longitude is therefore 60° west.

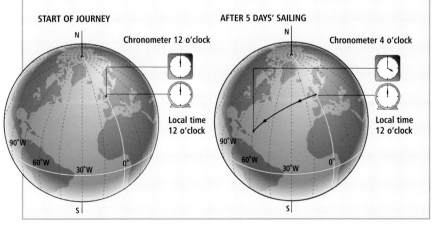

ASTRONOMY AND MATH

1750 French astronomer Guillaume Legentil de la Galaisière (1725–92) discovers the Trifid Nebula (M20) in the constellation Sagittarius. It is named by English astronomer John Herschel (1792–1871) for the three dark rifts that appear to divide the nebula and join at its center.

1751 French astronomer Nicolas de Lacaille (1713–62) leads an expedition to the Cape of Good Hope to observe the stars of the southern skies. His observations enable him to make the first accurate calculation of the distance between the Earth and the Moon.

1755 German philosopher Immanuel Kant (1724–1804) explains his theory for the formation of the solar system in his *Universal Natural History and Theory of the Heavens*. He postulates that it is created from a spinning gaseous nebula and that our Galaxy is just one of many in the Universe.

CHEMISTRY AND PHYSICS

1751 Swedish chemist Axel Cronstedt (1722–65) discovers nickel.

1752 American scientist and politician *Benjamin Franklin* (1706–90) demonstrates the electrical nature of lightning in his famous kite-flying experiment.

1758 German chemist Andreas Marggraf (1709–82) introduces flame tests as a method of chemical analysis. (Various metallic elements give a characteristic color to a gas flame.)

BIOLOGY AND MEDICINE

Robert Bakewell bred the Leicestershire sheep, a new breed that produced long, coarse wool and high-quality meat.

1752 French chemist René-Antoine Ferchault de Réaumur (1683–1757) discovers the part played by gastric juices in the digestion of foods.

1760s English agriculturist Robert Bakewell (1725–95) uses selective breeding to produce improved varieties of farm animals.

1761 Austrian physician Leopold Auenbrugger (1722–1809) introduces the diagnostic technique of percussion, which involves tapping the patient's chest and listening to the quality of the sound produced.

ENGINEERING AND INVENTION

1750 German engineer Johann Segner (1704–77) constructs a reaction waterwheel (in which the wheel is turned by the force of a jet of water).

1753 Scottish engineer Charles Morrison invents a 26-wire telegraph (one wire for each letter of the alphabet).

1756 English engineer **John Smeaton** (1724–92) invents "hydraulic lime" cement that sets underwater (for building lighthouses).

1757 English clockmaker Thomas Mudge (1715–94) devises a lever escapement for a watch, although it is not used until 1770.

1757 English engineer Henry Berry (1720–1812) completes the Sankey Brook Navigation, the first modern canal, in northwestern England.

See also Domestication of Animals **1**:24–25; Benjamin Franklin **4**:40–41; Textile Machines **5**:12–13; Canal Transportation **5**:20–21

1758 German astronomer Johann Palitzsch (1723–88) observes Halley's comet when it returns as predicted by English astronomer **Edmond Halley** (1656–1742) in 1682.

1760 German physicist Johann Lambert (1728–77) formulates Lambert's law, which states that the illuminance provided by light striking a surface at right angles is inversely proportional to the square of the distance between the surface and the light source.

1761 Italian physician Giovanni Morgagni (1682–1771) founds the science of pathology with his book *On the Seats and Causes of Disease... .*

1762 The flightless dodolike bird called the solitaire (*Pezophaps solitaria*) becomes extinct (on the Indian Ocean's Rodriguez Island).

1758 English cotton weaver Jedediah Strutt (1726–97) invents the stocking frame, a machine for making ribbed hosiery.

1758 English optician John Dollond (1706–61) makes achromatic lenses that consist of two pieces of different glass types (crown glass and flint glass).

1761 English engineer **James Brindley** (1716–72) completes the construction of the Bridgewater Canal, built for the duke of Bridgewater to carry coal from the mines to Manchester in the north of England.

1762 English nobleman John Montagu, earl of Sandwich (1718–92), invents the sandwich (so that he did not have to leave the gambling tables to take his meals).

1761 Russian scientist Mikhail Lomonosov (1711–65) observes a transit of Venus across the Sun's disk and deduces that Venus has an atmosphere.

1762 English astronomer James Bradley (1693–1762) produces a star catalog with measured positions of 60,000 stars.

1761 Scottish chemist Joseph Black (1728–99) introduces the concept of latent heat. Latent heat is the "extra" heat that must be added to an object at its melting point before it actually melts (or to a liquid at its boiling point before it boils).

1762 The École Nationale Vétérinaire is established in Lyons, France, the world's first national veterinarian college.

1763 German botanist Josef Kölreuter (1733–1806) discovers the role of insects in the pollination of flowers.

1764 English mechanic James Hargreaves (d. 1778) invents the spinning jenny for spinning many cotton or woolen threads at once.

1764 French mathematician Joseph Lagrange (1736–1813) explains the Moon's libration, the slight "nodding" of its axis that allows us to see more than 50 percent of its surface.

1762 English physicist John Canton (1718–72) demonstrates that contrary to previous beliefs, water is a compressible liquid.

Josef Kölreuter discovered in 1763 how insects pollinate flowers.

1764 French engineer Pierre Trésaguet (1716–96) introduces his system of road building to France.

The original spinning jenny of 1764 had an almost horizontal wheel, allowing up to 12 threads to be spun at once.

The Bridgewater Canal was completed in 1761.

ASTRONOMY AND MATH

CHEMISTRY AND PHYSICS

BIOLOGY AND MEDICINE

ENGINEERING AND INVENTION

1750–1764 A.D.

Benjamin Franklin (1706–90)

▲ An engraved print from 1789 shows a caricature of Benjamin Franklin melting the works of earlier authors in order to create his own pot of "anecdotes."

KEY DATES

1733 *Poor Richard's Almanac* first published

1742 Franklin stove

1752 Kite experiment and lightning rod

1784 Bifocal eyeglasses

▶ Franklin was the inventor of bifocal eyeglasses, improving both distance and near vision for the wearer.

*B*enjamin Franklin was a complex character, an unusual combination of statesman and scientist. He played an important part in the development of the young United States and made major discoveries in physics. He also was a talented inventor, and some of his inventions are still in use throughout the world today.

Born in Boston into a family of 17 children, Benjamin Franklin left school at the age of ten. Two years later he was apprenticed to his older brother James, a printer. When he was just 18, he took over publication of the *New England Courant*, a weekly newspaper founded by his brother. He did not stay for long; instead, he went to Philadelphia and worked as a printer himself. In 1724 he set sail for England. He returned home two years later and published the first volume of *Poor Richard's Almanac* in 1733, a collection of articles on a wide range of subjects to "convey instruction among the common people." He held various public offices and helped draft the Declaration of Independence in 1766. He traveled to France to raise help for the American cause in the War of Independence, and while in Paris he witnessed the **Montgolfier brothers'** first hot-air balloon flight in 1783. He was a staunch supporter of the abolition of slavery; he retired from public life in 1788.

During his lifetime he also conducted scientific experiments. The best known, in 1752, was one of the most dangerous experiments ever undertaken. He attached a metal key to the moistened string of a kite, which he flew during a thunderstorm. Electric "fluid" flowing down the string caused sparks to jump between the key and a Leyden jar (a primitive electrical condenser). Franklin had established the electrical nature of lightning and coined the words "positive" and "negative" to describe the two types of static electricity. Several European scientists who tried to repeat the experiment were struck by lightning and killed. Franklin, however, devised a means of protection: He invented the lightning rod, a pointed conductor located at the top of a building and connected to the ground by a thick wire attached to a plate buried in the soil. Today all tall buildings have lightning rods. He theorized that thunderclouds are electrically charged and recognized the aurora borealis (northern lights) to be electrical in nature.

"Nothing in life is certain but death and taxes."

On being asked of what use was the new hot-air balloon: *"What is the use of a new-born child?"*

Benjamin Franklin

Franklin had many other scientific interests. Unlike most of his contemporaries, he rejected **Newton**'s corpuscular theory of light (that light travels as particles), favoring the wave theories of **Robert Hooke** and others. He suggested that the rapid heating of air near warm ground causes it to expand and spiral upward, producing tornadoes and waterspouts. He investigated the course of the Gulf Stream, the current of warm water that flows across the Atlantic Ocean, and he suggested that ships' captains should use a thermometer to locate and benefit from the current (or avoid it, depending on the direction they were sailing). In 1824 the Franklin Institute was founded in Philadelphia in his honor.

▶ An artist's romantic depiction of Franklin's famous experiment. He flew a kite in a thunderstorm and found that the wet string conducted electricity from the storm to charge a capacitor.

Hearth and Home

At a more domestic level Franklin is credited with inventing the rocking chair and, in 1742, the Franklin stove. Also called the Pennsylvania fireplace, the stove burned efficiently because it was equipped with an underfloor draftpipe. Franklin was nearsighted and had to wear eyeglasses for reading. He also needed different eyeglasses for distance vision. Annoyed with having to change eyeglasses all the time, he invented bifocals in about 1784. They had split lenses—the upper half for distance vision and the lower half for near vision. And in a bid to save fuel when reading on dark evenings, he suggested the introduction of daylight saving time.

See also Isaac Newton **4:**20–21; Michael Faraday **5:**36–37

ASTRONOMY AND MATH

1765 Colonial American scientist John Winthrop (1714–79) includes calculations of comet masses in his book *Account of Some Fiery Meteors*. In 1772 he presents Harvard University with its first telescope.

1767 German physicist Johann Lambert (1728–77) proves that π (the ratio of a circle's circumference to its diameter) is an irrational number—it cannot be expressed as a fraction involving two whole numbers.

1767 English astronomer Nevil Maskelyne (1732–1811) edits the first edition of the *Nautical Almanac*, a book of astronomical and other tables that has been published every year since.

1767 Swiss mathematician Leonhard Euler (1707–83) sets out the rules of algebra in his book *Vollständige Anleitung zur Algebra* (Complete Instruction in Algebra).

1769 A transit of Venus takes place (when the planet Venus passes in front of the Sun's disk). Expeditions to observe it are made by English explorer James Cook (1728–79) to Tahiti and German naturalist Peter Pallas (1741–1811) to the border between China and Russia.

1770 Swedish astronomer Anders Lexell (1740–84) discovers Lexell's comet, the first of the so-called short-period comets to be found (at that time it orbited the Sun once every 5.6 years).

CHEMISTRY AND PHYSICS

1766 French chemist Pierre Macquer (1718–84) publishes the first systematic dictionary of chemistry.

1766 English scientist **Henry Cavendish** (1731–1810) identifies hydrogen, which he calls "inflammable air."

1769 English chemist **Joseph Priestley** (1733–1804) writes *The History and Present State of Electricity,* in which it is suggested that electrical forces follow an inverse square law.

1770 Swedish chemist **Karl Scheele** (1742–86) isolates tartaric acid from cream of tartar.

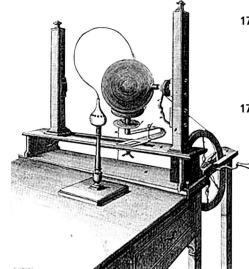

1771 Swiss geologist Jean Deluc (1727–1817) uses a barometer to find the heights of mountains (atmospheric pressure decreases with altitude).

1771 Italian physician and physicist Luigi Galvani (1737–98) shows that the muscles of a dissected frog twitch when they are stimulated by electricity; later he demonstrates that certain metals cause the same effect.

Left: Joseph Priestley's electrical apparatus.

Right: Luigi Galvani used animals such as frogs and sheep in his experiments with electricity in 1771.

ENGINEERING AND INVENTION

1765 English clockmaker Thomas Mudge (1715–94) invents the free-anchor escapement for clocks that uses an oscillating balance wheel.

1765 Italian naturalist Lazzaro Spallanzi (1729–99) devises a method of preserving food using hermetically sealed containers.

1765 The world's first mining academy is established in Freiberg, Germany.

1765 The first medical school in the United States is established at the College of Pennsylvania.

1765 Scottish engineer *James Watt* (1736–1819) builds a steam engine with a separate condenser.

1768 French chemist Antoine Baumé (1728–1804) invents a hydrometer (which he calls an aerometer) and a new density scale for graduating it.

c.1769 English engineer **John Smeaton** (1724–92) designs a cylinder-boring machine.

1769 English manufacturer Richard Arkwright (1732–92) produces the spinning frame for spinning strong cotton thread.

1769 French military engineer Nicolas-Joseph Cugnot (1725–1804) builds a steam-powered three-wheel vehicle (for towing guns).

Cugnot's steam-powered vehicle of 1769 was the first true automobile, capable of carrying four passengers at speeds of up to 2.25 miles (3.6 km) per hour.

See also The Steam Engine **4**:44–45; James Watt **5**:8–9; Textile Machines **5**:12–13

1772 German astronomer Johann Bode (1747–1826) propounds a numerical relationship between the distances of the planets and the Sun, known as Bode's law. It was first formulated in 1766 by German astronomer Johann Titius (1729–96) but is now thought to be simply a coincidence.

1773 French mathematician Pierre-Simon de Laplace (1749–1827) studies the orbits of the planets and uses Newton's law of gravity to prove that the solar system is inherently stable.

1773 German-born English astronomer **William Herschel** (1738–1822) studies the proper motions of 13 distant stars and figures out that the Sun is gradually moving through space toward the constellation Hercules. A year later he begins making regular observations using his new reflecting telescope, which has a focal length of 5.6 feet (1.7 m).

1774 German astronomer Johann Bode (1747–1826) founds *Astronomisches Jahrbuch* (Astronomical Yearbook), the German equivalent of the *Nautical Almanac.*

1774 English astronomer Nevil Maskelyne (1732–1811) determines the gravitational constant (the constant that appears in Newton's law of gravitation) and is thus able to calculate the average density of the Earth.

1772 French mineralogist Jean Romé de Lisle (1736–90) publishes *Essai de Cristallographie* (Essay on Crystallography) in which he states that crystals have constant angles between their faces.

1772 Scottish chemist Daniel Rutherford (1749–1819) discovers nitrogen.

1772 Swedish chemist **Karl Scheele** (1742–86) discovers oxygen but does not publish his findings until 1777.

1774 English chemist **Joseph Priestley** (1733–1804) identifies oxygen and publishes his findings. In the same year he describes ammonia.

1774 Swedish chemist **Karl Scheele** (1742–86) discovers barium, chlorine, and manganese.

1774 Swedish mineralogist Johan Gahn (1745–1818) is the first to isolate manganese.

1774 French chemist **Antoine Lavoisier** (1743–94) demonstrates the law of conservation of mass (in a chemical reaction).

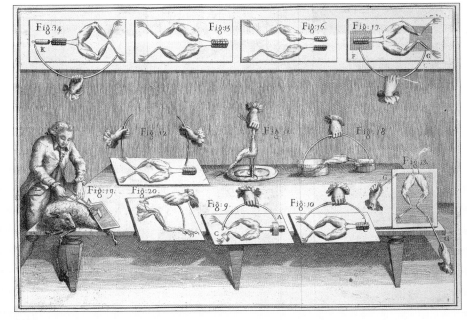

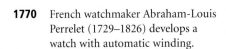

1769 English manufacturer Eleanor Coade (1733–1821) introduces a type of artificial stone (known as coade stone) for making molded sculpture and decorative objects.

1769 American astronomer David Rittenhouse (1732–96) constructs the first astronomical telescope in the U. S.

1770 English chemist **Joseph Priestley** (1733–1804) invents the pencil eraser.

1770 French inventor Jacques de Vaucanson (1709–82) constructs textile machinery that incorporates a chain drive, the first to be employed in Europe.

1770 French watchmaker Abraham-Louis Perrelet (1729–1826) develops a watch with automatic winding.

1772 French engineer Jean Perronet (1708–94) completes the Pont de Neuilly stone bridge over the Seine River near Paris. It becomes a model for later bridges.

1772 English inventor Henry Clay patents a method for making papier mâché.

1774 English instrument maker Jesse Ramsden (1735–1800) produces a "dividing engine" for engraving calibrations on scientific instruments.

1774 English engineer John Wilkinson (1728–1808) patents a precision cannon-boring machine, also used for making cylinders for steam engines.

1765–1774 A.D.

THE STEAM ENGINE

From the time of the ancient Egyptians until the end of the 17th century only wind and water power provided an alternative to the muscles of animals and humans. But the situation changed dramatically with the invention of the steam engine, which involved a sequence of events beginning in 1690 and culminating in 1765 with the work of James Watt.

▲ Coal mines soon made use of the new steam engines to run the pumps that removed water from deep underground.

The earliest steam engines did not use the pressure of steam to provide power. Instead, they used the pressure of the air. For that reason they are more accurately called atmospheric engines. The first was devised by **Denis Papin** (1647–1712), a French physicist working in England. A vertical, open-ended cylinder with a close-fitting piston had water inside its base. A fire heated the base of the cylinder, causing the water to boil and turn into steam. Steam pressure lifted the piston, which remained raised while the cylinder cooled. The steam condensed back to liquid water, creating a partial vacuum in the cylinder. Then atmospheric pressure on the upper end of the piston pushed it down again. A rope connected to the piston and moving over a pulley could be used to lift a load or work a pump.

A similar arrangement, patented in 1698 by English mining engineer **Thomas Savery** (*c.*1650–1715), made the atmospheric engine into a practical steam pump. It had no piston or other moving parts, just hand-operated valves to provide continuous operation. Steam from a boiler passed into a working chamber that was sprayed with cold water to condense the steam. The partial vacuum that was created as a result lifted water through a one-way valve into the chamber. Steam was then let in again, which forced the water out and up through another one-way valve.

▶ This model of Savery's atmospheric engine dates from the 1750s, by which time Newcomen's steam engine had taken over.

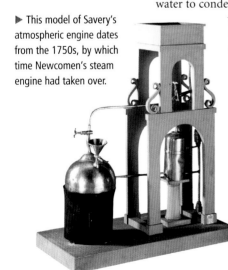

In 1712 English engineer **Thomas Newcomen** (1663–1729) perfected the first engine to use steam pressure to work a piston. Because of the way they were constructed, people generally referred to Newcomen's engines as beam engines. Newcomen could not patent his engine because its principle was too close to that of Thomas Savery's, so the two men went into partnership and built engines together.

This was the stage that steam power had reached in 1764, when Scottish engineer *James Watt* (1736–1819) received a model of a Newcomen engine to repair. He realized how much energy is wasted by first heating the cylinder and then cooling it. In 1765 he hit on the idea of adding a separate external condenser. In addition he used steam to push up the piston and then—by admitting low-pressure steam on the other side—to push it down again. This double action greatly improved the efficiency of the machine.

The next development was an increase in the pressure of steam in the engine—in the terminology of the day, "strong steam." This had to await the introduction of improved cylinders with better-fitting pistons and boilers that were safe at high steam pressures. They became available by 1800, when Watt's master patent expired. The following year English inventor **Richard Trevithick** (1771–1833) started building double-action, high-pressure engines. Trevithick removed the separate condenser and used

Newcomen's Steam Engine

The Newcomen engine was the first genuine steam engine. Unlike Savery's engine, it used steam pressure to push the piston directly, not just atmospheric pressure. Even so, the engines were similar enough for Newcomen to do a deal with Savery so that he could market engines under Savery's patent.

Steam from a boiler forced a piston up an open-ended cylinder, and then cold water sprayed into the cylinder condensed the steam, creating a partial vacuum, which sucked the piston down again. The piston rod connected to one end of a long beam; the other end of the beam connected to a pump. As the piston went up and down, the beam rocked and worked the pump continuously.

A Newcomen engine could produce about 12 strokes a minute. But because it needed to heat the water and cool the steam at every stroke, it used a lot of fuel. In 1765 Watt adapted Newcomen's engine by adding an external condenser. This improvement ensured that the cylinder stayed hot and the condenser remained cool all the time, saving significant amounts of fuel.

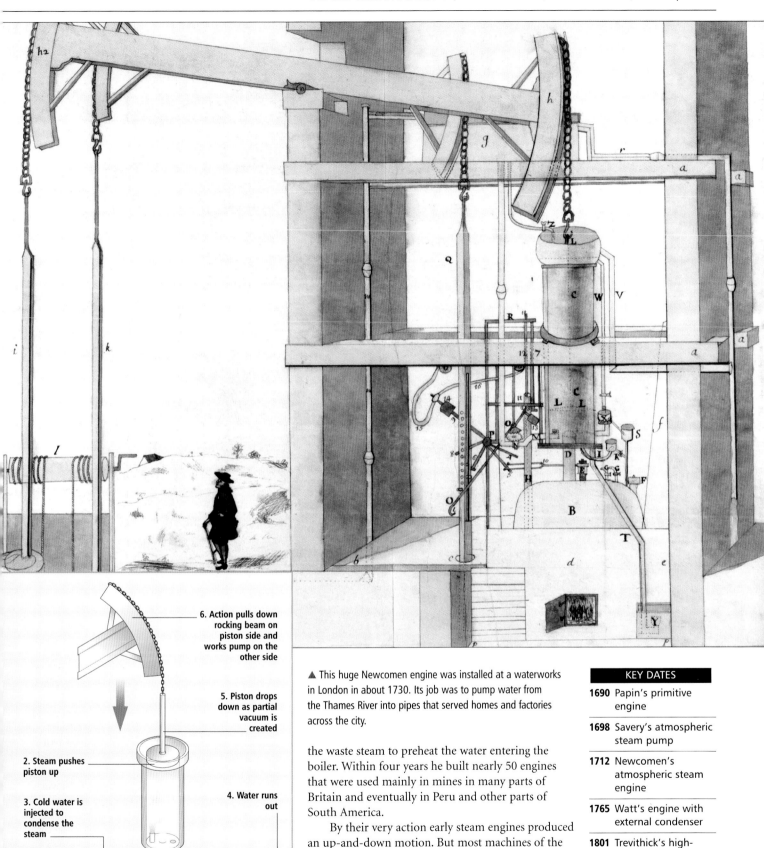

6. Action pulls down rocking beam on piston side and works pump on the other side

5. Piston drops down as partial vacuum is created

2. Steam pushes piston up

3. Cold water is injected to condense the steam

4. Water runs out

1. Water is heated and boiler produces steam

▲ This huge Newcomen engine was installed at a waterworks in London in about 1730. Its job was to pump water from the Thames River into pipes that served homes and factories across the city.

the waste steam to preheat the water entering the boiler. Within four years he built nearly 50 engines that were used mainly in mines in many parts of Britain and eventually in Peru and other parts of South America.

By their very action early steam engines produced an up-and-down motion. But most machines of the time, except pumps, required rotary motion. Until the advent of the steam engine most of them had been driven by waterwheels. Then in 1781 James Watt invented the sun-and-planet gear to make his engines provide a rotary final drive.

KEY DATES
1690 Papin's primitive engine
1698 Savery's atmospheric steam pump
1712 Newcomen's atmospheric steam engine
1765 Watt's engine with external condenser
1801 Trevithick's high-pressure engine

KEY PEOPLE 1625–1774 A.D.

*People whose names appear in **bold type** have their own articles in this section or in the "Key People" section of another volume. Names in **bold italics** indicate they are the subject of a special feature.*

Boyle, Robert (1627–91)

Robert Boyle was an Irish chemist and physicist who is remembered for his work on gases. He was born in Munster, Ireland, a son of the earl of Cork. From 1635 he was educated at Eton College in England, and three years later he went to study in Geneva, Switzerland. After returning to England in 1644, he lived on his estate in Dorset in the south of England, where he began studying "natural philosophy," i.e., science. Ten years later he established a laboratory in Oxford. Assisted by the versatile scientist **Robert Hooke** (1635–1703), he began experimenting with gases using the new vacuum pump designed in 1654 by German physicist Otto von Guericke (1602–86).

Boyle used the pump to increase air pressure in a container or to remove some of the air and so decrease the pressure. As a result of these experiments, in 1662 he put forward Boyle's law. The law states that at constant temperature the pressure of a gas is inversely proportional to its volume. In France and the rest of continental Europe it became known as Mariotte's law after French physicist Edmé Mariotte (1620–84), who discovered the law independently in 1676.

Brindley, James (1716–72)

James Brindley was an English engineer who built some of Britain's first canals. He was born in Thornsett in central England and lacked schooling. He could not read or write but nevertheless trained as an engineer after becoming apprentice to a millwright in 1733. He later set up in business on his own and worked on pumps for mines, making a "water engine" for draining a coal mine in 1752. In 1759 Francis Egerton, duke of Bridgewater, employed him to build a canal between the duke's coal mines in Worsley and Manchester. Brindley built a winding contour canal on one level with no locks. It was completed in 1761 and was known as the Bridgewater

Canal. Brindley went on to build or survey several other major canals, most of the work being done by his assistants while he hurried from contract to contract to check the work.

Cassini, Giovanni (1625–1712)

Giovanni Domenico Cassini was an Italian-born French astronomer, most famous for his studies of the planets and their moons. Born in Perinaldo, Republic of Genoa, he became professor of astronomy at the University of Bologna in 1650. In 1669 he was invited by King Louis XIV of France to take charge of the new Paris Observatory. He became a French citizen in 1673.

In 1668 he compiled a table of the movements of the four moons of Jupiter that *Galileo* (1564–1642) had discovered in 1610; this was the table used in 1676 by Danish astronomer Ole Rømer (1644–1710) when calculating the speed of light. Beginning in 1671, Cassini discovered four more of Jupiter's moons, and in 1675 he first noticed the gap that divides Saturn's rings into two parts. Since then the gap has been known as the Cassini division. He also calculated the distance to Mars, and from this figured out the distance between the Earth and the Sun (1 astronomical unit). His calculation, 87 million miles (140 million km), is 7 percent too small, but gave for the first time a reasonable idea of the size of the solar system. After his death his son Jacques (1677–1756) succeeded him at the Paris Observatory, and later the post went to his grandson César (1714–84).

Darby, Abraham (c.1678–1717)

Abraham Darby was an English iron founder who revolutionized the production of iron by developing a smelting process that used coke. Previously charcoal had been used in blast furnaces, but the fuel was expensive (and in short supply) and not strong enough to bear the weight of a large amount of iron ore. Coal had been tried but contained too much sulfur for the production of high-quality iron.

Darby founded the Bristol Iron Company in 1708. He moved his operation to Coalbrookdale on the Severn River, where there was a plentiful supply of iron ore and

coal to make coke. In 1709 he set up his blast furnace, which produced nearly pure iron that could be cast to make pots and other containers as well as decorative ironwork for buildings. It was later used to cast boilers and cylinders for the steam engines constructed by **Thomas Newcomen** (1663–1729). In turn, steam engines replaced the waterwheels previously used to power the bellows that produced the air blast for the furnaces.

Darby's family continued the business. Abraham Darby II (1711–63) produced steam boilers for **Richard Trevithick** (1771–1833) and cast-iron rails for his new railroads. Abraham Darby III (1750–89) cast the components for the first bridge to be built entirely of iron. It was completed in 1779 over the Severn at Coalbrookdale and was in continuous use until 1934, when it was closed to motorized traffic. It remained in use as a footbridge and still stands today.

Descartes, René (1596–1650)

René Descartes was a French mathematician and philosopher, one of the greatest thinkers of all time. He was born near Tours, and from 1604 he was educated for ten years at a Jesuit college. He graduated in law from the University of Poitiers in 1616 and spent the next ten years in military service and traveling throughout Europe. In 1628 he settled in Holland. He went to work as tutor to Swedish Queen Christina in 1649; while living there he nursed the French ambassador but caught pneumonia from him and died.

In 1619 Descartes concluded that the position of a point on a plane can be defined by a pair of coordinates that locate its distance from a pair of straight lines or axes (generally at right angles to each other). Now called Cartesian coordinates, they led Descartes to invent in 1637 coordinate geometry, sometimes called analytic geometry, in which lines and curves can be expressed in terms of algebraic equations. In other words, the various techniques of algebra can be applied to geometry.

In astronomy he tried to explain the motions of the planets in terms of vortexes swirling around the Sun. These ideas were later shown to be false by English scientist *Isaac Newton* (1642–1727).

Franklin, Benjamin (1706–90) *See* feature Vol. 4, pages 40–41.

Galilei, Galileo (1564–1642) *See* feature Vol. 4, pages 8–9.

Gassendi, Pierre (1592–1655)

Pierre Gassendi was the French astronomer and mathematician who made the first observation of a transit of the planet Mercury across the Sun's disk. He was born in Champtercier, was awarded a doctorate in theology from Avignon in 1616, and was ordained as a priest a year later, before obtaining the post of professor of philosophy at Aix. In 1645 he became professor of mathematics at the Collège Royal in Paris, where he met *Galileo* (1564–1642) and German astronomer **Johannes Kepler** (1571–1630). It was Kepler who predicted the time of Mercury's transit, which duly took place in 1631. Gassendi projected an image of the Sun from the eyepiece of his telescope to avoid the danger of having to look directly at the Sun. He also studied eclipses and comets, and coined the name "aurora borealis" for the northern lights.

Gregory, James (1638–75)

James Gregory was a Scottish mathematician who also made contributions to astronomy and physics. He was born in Drumoak in the northeast of Scotland. In 1662 he graduated from Aberdeen University and went to London. In 1663 he published a description of a reflecting telescope (now known as a Gregorian telescope), which he had conceived two years earlier. In this design the main mirror reflects light to a second small, curved mirror, which in turn reflects the light back through a hole in the center of the main mirror to an eyepiece. But it is unlikely that there was anyone skilled enough to make such an instrument at that time. He also used a bird's feather as a diffraction grating to project the Sun's spectrum onto the wall of a darkened room.

He lived in Padua, Italy, from 1664 to 1667 and published a book on the geometry of the circle and the hyperbola. He was appointed professor of mathematics at St. Andrews University in 1668 and moved to a similar post at Edinburgh University six years later. He introduced the term "infinite series" to mathematics, distinguishing between convergent and divergent series, and they were the subjects of most of his later studies.

Halley, Edmond (1656–1742)

Edmond Halley was an English astronomer best known for the comet that bears his name. He was born in London and educated there and at Oxford University. While still a student he published several papers on the planets and in 1676, before completing his studies, went to St. Helena in the southern Atlantic to compile a catalog of Southern Hemisphere stars. He joined **Giovanni Cassini** (1625–1712) at the Paris Observatory in 1680, where he calculated the orbits of more than 20 comets. He showed that comets that had been seen in 1531, 1607, and 1682 were, in fact, a single comet and predicted its return in 1758. When it returned on time (on December 25), it was named Halley's comet.

As well as his astronomical studies, Halley made contributions to physics, using **Boyle**'s law to correlate atmospheric pressure with heights above sea level. In 1686 he investigated monsoons and the trade winds, studied magnetic variation (caused by the slow movement of the north magnetic pole), and was instrumental in helping *Isaac Newton* (1642–1727) publish *Principia* in 1687—in fact, he read the proofs and financed the printing. In 1693 he published *Breslau Tables of Mortality*, which actuaries used to calculate premiums for annuities and life insurance. He designed an improved diving bell and an accurate barometer. In 1720 he succeeded John Flamsteed (1646–1719) to become the second Astronomer Royal, a post he held for 21 years.

Harvey, William (1578–1657)

William Harvey was an English physician who discovered the circulation of the blood. He was born in southern England, the son of a farmer. He went to school in Canterbury before attending Cambridge University, graduating in medicine in 1597. After working under Italian physician Girolamo Fabrizio (1537–1619) in Padua, Italy, for five years, he opened a medical practice in London. He became Physician Extraordinary to King James I and later accompanied King Charles I on his campaigns during the Civil War. He retired to private life in 1646.

Harvey's great discovery was published in 1628. He announced that "the blood performs a kind of circular motion" through the body. From the time of Greek physician **Galen** (*c*.129–*c*.200) people believed that the blood was formed in the liver and moved from there throughout the body. Arteries and veins were supposed to carry different substances. Harvey noted valves in the veins that allow blood to flow toward the heart (but not away from it). The heart, he argued, pumps blood away, around the body in arteries, into the veins, and back to the heart. He tested his theory by dissecting animals and found that the blood travels from the right side of the heart and through the lungs before returning to the left side of the heart for distribution to the body. He could not explain how blood got from the arteries to the veins; this had to await the discovery of capillary blood vessels by English physician Henry Power (1623–68) in 1649 (confirmed by Italian anatomist Marcello Malpighi, 1628–94, in 1661).

Hooke, Robert (1635–1703)

Robert Hooke was an English scientist who made major contributions to astronomy, physics, and engineering. The son of a clergyman, he was educated at Westminster College, London, and Oxford University, where he assisted **Robert Boyle** (1627–91) with his experiments on gases. In astronomy Hooke discovered Jupiter's Great Red Spot and in the same year (1664) proposed that the planets are held in their orbits by gravity. *Isaac Newton* (1642–1727) had the idea at about the same time. This led to one of many disputes he was to have with Newton. Hooke did not express his theory in mathematical terms—as an inverse square law of gravity— until 1678, and the following year he outlined his theories on planetary motion in a letter to Newton. Newton did not publish his *Principia* until 1687, and Hooke therefore claimed the discovery as his own. Hooke also built one of the first reflecting telescopes—another area in which Newton was active—following the 1663 design of Scottish astronomer **James Gregory** (1638–75).

Also in the 1660s Hooke began studying natural objects using an improved compound microscope he constructed. He looked at microscopic animals and at plant cells, which he likened to "little boxes," and coined the word "cell" to describe them. He published his findings in 1665 in a beautifully illustrated book, *Micrographia*. The use of the microscope prompted him to consider the nature of light. He concluded it behaved like waves. Once again his ideas conflicted with Newton's. The latter championed the corpuscular theory (that light travels as tiny particles).

Less contentious was Hooke's major contribution to physics: the law that bears his name. Postulated in 1676, Hooke's law states that when an elastic object stretches, stress is proportional to strain. Stress is force per unit area causing the stretching, and strain is the accompanying change in dimensions of the object. In scientific terms anything that stretches is elastic. The stretching can be obvious, as in rubber, or much less noticeable, as with most metals. He also showed that most materials expand on heating.

Hooke applied his new knowledge of the properties of materials in various ways. He had already (1658) invented the hairspring as part of the balance-wheel mechanism in a watch. The coiled hairspring oscillates as it repeatedly tightens and slackens to control the gradual release of driving power from the mainspring. Even this development involved a priority dispute with Dutch scientist **Christiaan Huygens** (1629–95). In 1660 Hooke invented the anchor escapement, which exerts a similar control in a pendulum clock. In 1667 his contribution to science was an anemometer, an instrument for measuring wind speed. And in 1676 he invented the universal joint, a device that couples a rotating shaft to another shaft aligned at an angle. No longer did the source of power, such as an engine, have to be precisely in line with the final drive. A universal joint is still an important part of the transmission of every truck and car on the road.

Huygens, Christiaan (1629–95)

Christiaan Huygens was a Dutch scientist. Rated second only to *Isaac Newton* (1642–1727) for his contributions to knowledge, Huygens made important discoveries in astronomy, mathematics, and physics. He was born in The Hague, son of a famous Protestant poet, and studied at the

University of Leyden (now Leiden) and at Breda. He went to Paris and became a founding member of the Academy of Sciences, where he remained from 1666 until 1681, when he returned to Holland to escape from religious intolerance.

In 1655, using a home-made refracting telescope, he observed the rings of Saturn and discovered Titan, the planet's largest moon. A year later he put into practice an idea of *Galileo* (1564–1642) and designed a pendulum clock, which was finally made in 1657 by Dutch clockmaker Salomon Coster. Huygens described the clock, together with theories about the pendulum and other ideas in mechanics, in his book *Horologium Oscillatorium* (The Pendulum Clock) of 1673.

Huygens went on to study light. He subscribed to the view that space is filled with an invisible "ether" through which light travels as waves spreading out from a source in the same way as ripples on a pond's surface. Each point on a wave (ripple) can act as a source of secondary waves. Using this theory, Huygens satisfactorily explained the phenomena of reflection and refraction. He also predicted that light travels more slowly in a denser medium than in a less dense one, a phenomenon that could not be explained by Newton's alternative corpuscular theory. Eventually the wave theory gained favor, but not until it was championed in 1807 by English physicist Thomas Young (1773–1829). The elusive ether, however, proved to be nonexistent.

Kepler, Johannes (1571–1630)

Johannes Keppler was a German astronomer famous for proposing three laws of planetary motion. He was born Johann von Kappel near Stuttgart in southwest Germany, son of a mercenary soldier who died five years later. Kepler graduated from the University of Tübingen in 1591 with a degree in theology. Three years later he became professor of mathematics at the Lutheran school in Graz, where he published an annual almanac of astrological predictions. In 1600 he went to Prague to join Danish astronomer Tycho Brahe (1546–1601). On the latter's death he took over as imperial mathematician to Holy Roman Emperor Rudolph II. In 1604 he discovered a nova ("Kepler's star") in the constellation Ophiuchus. From his studies of

Mars he concluded that the planets do not orbit the Sun in perfect circles but in ellipses with the Sun at one focus (Kepler's first law), and that a line from the focus to the orbit sweeps out an equal area of space in equal amounts of time (Kepler's second law). Kepler's third law (1619) relates a planet's period of revolution to its average distance from the Sun. In 1627 he published the *Rudolphine Tables*, containing tables of planetary motions and a catalog of 1,005 stars based on Tycho's observations. They were used by astronomers until the 18th century.

Lavoisier, Antoine (1743–94)

Antoine-Laurent Lavoisier was a French scientist and the founder of modern chemistry. He was born in Paris to prosperous parents and educated at the Collège Mazarin. He qualified as a lawyer in 1764, having studied various branches of science in his spare time. He assisted for three years in drawing up a geological map of France and was elected to the Royal Academy of Science in 1768. In the same year he became a member of a tax farm (a private agency that collected government taxes), rising to become farmer-general. During a period of the French Revolution known as the Reign of Terror (1793–94) Lavoisier and 27 other tax farmers were arrested and charged with offences against the revolution. In May 1794 all 28 were tried, convicted, guillotined, and buried in unmarked graves.

Lavoisier's early chemical experiments concerned combustion. In 1772 he burned a diamond to produce "fixed air" (carbon dioxide) and showed that phosphorus and sulfur gained weight when they burned because they combined with air. He said that the weight gain of calcined (strongly heated) metals occurred for the same reason. In 1774 English chemist **Joseph Priestley** (1733–1804) told Lavoisier of a new "air" produced by heating mercuric oxide. The gas vigorously supported combustion, so Lavoisier repeated the experiment and identified the gas, which he named oxygine, with the component of air used up during combustion. Chemists previously believed that when a substance burns, it gives up the mysterious substance "phlogiston" (and should therefore lose weight). Lavoisier showed that a burning substance combines with oxygen and gains weight. This proved to be the death knell of the phlogiston theory. In 1783 he showed that water is a compound of hydrogen and oxygen (and not an element), although English scientist **Henry Cavendish** (1731–1810) had already anticipated this. In 1787 Lavoisier published a book, together with three other French chemists, outlining a new systematic way of naming chemicals. Two years later he wrote a chemistry textbook, listing the 33 elements then known and outlining how chemistry was to develop as a science.

Lavoisier also noted that combustion and respiration are similar in many respects and from 1785 he carried out many experiments on breathing using small animals. He went on to demonstrate that as in combustion of a fuel, breathing produces carbon dioxide, and oxygen is essential to life. In 1790 he joined the commission that recommended the adoption of the metric system in France.

Leeuwenhoek, Antonie van (1632–1723)

Antonie van Leeuwenhoek was a Dutch amateur scientist who is remembered as the maker of the first simple microscopes. Son of a basketmaker, he was born in Delft in southwest Netherlands, and received little formal education before being apprenticed to a linen draper in 1648. About six years later he established his own business and from 1660 was a chamberlain in Delft's law courts.

Leeuwenhoek developed a skill for grinding lenses (for inspecting cloth) and from about 1670 used them to make simple microscopes that magnified up to 200 times. Each microscope (he made nearly 250) had only a single, small lens but was powerful enough to reveal the existence of microscopic creatures ("animalcules"), including protozoa in pond water, bacteria in the tartar from his teeth, and spermatozoa in the semen of dogs and humans. He also investigated lice, aphids, ants, bees, and other insects, and discovered blood cells (corpuscles), nerve structures, and blood capillaries.

Leibniz, Gottfried (1646–1716)

Gottfried Wilhelm Leibniz was a German mathematician and philosopher who introduced binary arithmetic. He was born in Leipzig, son of a Lutheran professor of moral philosophy, and educated at the universities there and in Jena and Altdorf. After graduating in 1666, he joined the staff of the elector of Mainz and spent three years in Paris. There and on trips to London he met with scientists **Robert Boyle** (1627–91) and **Christiaan Huygens** (1629–95). In 1677 he went to work for the elector of Hanover, where he remained for almost 40 years.

Among Leibniz's many contributions to mathematics was the discovery of infinitesimal calculus. He published his results in 1684. *Isaac Newton* (1642–1727) claimed the discovery as his own; but even though he had completed the work earlier, he did not publish until 1687. The two men arrived at their results independently, but the dispute caused bitter rivalry between them. In 1679 Leibniz introduced binary notation (numbers to the base 2), which uses only two digits: 1 and 0. It is the basis of all modern digital computers. He designed his own calculating machine, which was finally built nearly 80 years after his death.

Linnaeus, Carolus (1707–78)

Carolus Linnaeus was a Swedish naturalist who devised the binomial (two-name) system of classifying plants and animals and founded the science of taxonomy. Born Carl von Linné in Råshult, southern Sweden, he was the son of a pastor. He studied medicine and then botany at universities in Lund and Uppsala. His main interest was in plants: Between 1732 and 1735 he traveled around Lapland and other European countries and found 100 new species. He settled in Holland and qualified as a doctor of medicine. In 1735 he published *Systema Naturae* (Systems of Nature), in which he classified objects into three kingdoms: animal, plant, and mineral. He placed flowering plants into classes based on the characteristics of their stamens and subdivided the classes into orders based on their number of pistils.

Linnaeus introduced binomial naming in 1749, and in his book *Species Plantarum* (Species of Plants) of 1753 he gave each plant a Latinized genus name and a species name. For example, using his system, the American beech tree is *Fagus grandifolia* (genus *Fagus*, species *grandifolia*). The modern convention is for the names to be printed in italic letters with a capital letter only for the genus. The names of orders and classes are not printed in italics. Linnaeus practiced as a physician after his return to Sweden in 1738, before becoming a professor of medicine and botany at Uppsala University in 1741.

Newcomen, Thomas (1663–1729)

Thomas Newcomen was an English engineer. Beginning in 1705, he invented an atmospheric steam engine. An improvement on the 1698 steam pump of **Thomas Savery** (*c*.1650–1715), it paved the way for the later contributions of **Richard Trevithick** (1771–1833) and *James Watt* (1736–1819), which in turn helped make possible the Industrial Revolution.

Newcomen was originally a blacksmith and toolmaker from Dartmouth in southwest England. Together with a plumber named John Calley (or Cawley) he set out to design a steam pump to replace the horse-powered pumps used for draining water from mines. By 1712 they had built such an engine.

The original Newcomen engine had a vertical brass cylinder, open at the top and containing a piston. Low-pressure steam let in below the piston forced it upward to rock a long, pivoted crossbeam with pump rods at its other end. At the top of the stroke a jet of cold water was sprayed into the cylinder to condense the steam, creating a partial vacuum. As a result, atmospheric pressure forced the piston down again, while at the same time lifting the pump rods. The cycle was repeated, and valves operated by a second rod let in steam or water to the cylinder as required. The engine generated about 5.5 horsepower (4,100 watts). The machine was similar enough to Savery's for the two men to go into partnership making the new pumping engine. When Savery died in 1715, the patent rights went to a syndicate. It continued to exploit the invention, which used a cylinder of iron instead of brass. It was in common use in tin mines and elsewhere in Britain by 1725.

Newton, Isaac (1642–1727) See feature Vol. 4, pages 20–21.

Papin, Denis (1647–1712)

Denis Papin was a French mathematician and physicist but is best known for inventing a "steam digester"—the first pressure cooker.

Papin was born in Blois in northern France, the son of a doctor. He was educated mostly by his father before going to study medicine at the University of Angers. In 1673 he went to Paris, where he became assistant to Dutch scientist **Christiaan Huygens** (1629–95) working on gases. Two years later he was in London doing similar work with **Robert Boyle** (1627–91). He made his pressure cooker in 1680 and then went to Venice for three years as a demonstrator of experiments for the scientific academy. From 1684 he was back in London as curator of experiments for the Royal Society and finally settled in Marburg (Maribor), Germany, in 1687 as professor of mathematics. In 1690 he described a primitive steam engine and in 1707 proposed modifications to the steam pump of **Thomas Savery** (c.1650–1715).

Pascal, Blaise (1623–62)

Blaise Pascal was a French mathematician, physicist, and philosopher. He was born in Clermont-Ferrand in southern France, son of a mathematician and government official. He moved to Paris in 1630 after his mother died and was educated by his father. When only 11 years old, he figured out for himself **Euclid**'s first 23 theorems, and five years later he published an essay on mathematics. It was so advanced that French mathematician **René Descartes** (1596–1650) refused to believe it was the work of a person so young.

In 1641 he moved to Rouen and while there he made a wooden calculating machine to help in his father's work. In 1647 he repeated the earlier experiments of Italian physicist **Evangelista Torricelli** (1608–47) with a mercury barometer. By taking one of these instruments to the top of the Puy de Dôme, a mountain over 3,900 feet (1,200 m) high, he confirmed that atmospheric pressure decreases with altitude. In the same year, to test some of his ideas on probability, it is thought that he invented a primitive version of the roulette wheel.

Pascal's interest in pressure led him to investigate liquids. He concluded that within the body of a liquid the pressure is everywhere the same (Pascal's law), and that pressure applied at one place spreads immediately throughout the liquid. This is the principle of the hydraulic press and other hydraulic machines, including the syringe.

In 1654 Pascal had a religious revelation and thereafter largely gave up mathematics. He entered a Jansenist retreat and concentrated on religious studies. His name lives on in two modern applications. The SI unit of pressure is called the pascal (equal to 1 newton per square meter), and a high-level computer programing language developed by Swiss-born programmer Niklaus Wirth (1934–) in the 1960s is also called Pascal.

Priestley, Joseph (1733–1804)

Joseph Priestley was an English chemist best known for his work on gases, including the discoveries of oxygen and ammonia. He was born in Leeds, son of a weaver. His mother died when he was seven years old, and his Calvinist (nonconformist) aunt raised him. In 1752 he went to a seminary to train for the ministry and in 1755 became a minister, first in southeast England and then in Nantwich in the northwest. He began teaching languages at Warrington Academy in 1761. A year later he married Mary Wilkinson, sister of engineer John Wilkinson (1728–1808).

On a visit to London he met with **Benjamin Franklin** (1706–90), who stimulated Priestley's interest in electricity. Returning to Leeds in 1767, he lived near a brewery and experimented with the "fixed air" (carbon dioxide) that the brewing process yields in copious amounts. This work led to the quantity production of soda water (club soda), which soon became popular as a drink. Also at this time he wrote pamphlets that were critical of the government's attitude to the American colonies.

Priestley moved to Birmingham in 1780, where he met several other scientists and engineers (including Matthew Boulton, **James Watt**, Josiah Wedgwood, and Erasmus Darwin) at the Lunar Society. But his revolutionary political ideas angered local people. In 1791 some of them broke into his house and destroyed all his papers and apparatus. He escaped to London, but three years later he moved permanently to the United States and continued his scientific and theological work in Pennsylvania.

Priestley's approach to the study of gases was to heat things (or mixtures of things) and see what happened. When he started his work, chemists recognized only three gases: air (not yet known to be a mixture),

hydrogen, and carbon dioxide. To them he added ten more, including oxygen, ammonia, hydrogen chloride, sulfur dioxide, hydrogen sulfide, carbon monoxide, and nitrous oxide (laughing gas).

His best-known discovery, oxygen, was made in 1774. Using a magnifying glass to focus the Sun's rays, he heated mercury with air in a closed vessel to form an orange "crust" of mercuric oxide. When he heated the oxide on its own, it yielded an odorless, colorless gas in which a lighted candle burned much more brightly. This gas was oxygen (later named by **Antoine Lavoisier**, 1743–94). Priestley called it "dephlogisticated air." He did not recognize it as an element. That was left to Swedish chemist **Karl Scheele** (1742–86), who had already isolated it two years earlier but failed to publish his findings. Priestley never accepted that Lavoisier had disproved the phlogiston theory.

Savery, Thomas (c.1650–1715)

Thomas Savery was probably born in southwestern England, though little is known of the first 40 years of his life. From the middle 1690s he began taking out various patents, including one in 1694 for a machine to polish plate glass. In 1697 he designed a ship with paddle wheels driven by a (man-powered) capstan, and in 1707 he patented a floating mill whose waterwheel kept turning even when the water level rose and fell. He invented his machine for "raising water using fire" in 1698. It was a pump with a chamber that was first filled with steam and then cooled by pouring cold water over it. This created a partial vacuum that drew water into the chamber, forcing it out when the next lot of steam entered. His extended patent of 1699 overlapped with the idea for the atmospheric engine invented by **Thomas Newcomen** (1663–1729). The two men therefore joined forces and produced engines together.

Scheele, Karl (1742–86)

Karl Scheele was a Swedish chemist who, in a short career, discovered six chemical elements and many new compounds. He was born to a poor family in Stralsund (now in Germany). After an elementary education he was apprenticed to an apothecary in Göteburg (Gothenburg) in 1757. In 1770 he moved to

Uppsala, where he met with Swedish chemist Torbern Bergman (1735–84), who greatly encouraged him in his work. His final move was in 1775: He went to take charge of the pharmacy in the small town of Köping, where he remained for the rest of his short life.

Scheele's approach to chemistry was generally one of random experiment. In 1772 he discovered "fire air" (oxygen) but did not publish his results until 1777, and **Joseph Priestley** (1733–1804) usually gets the credit for this discovery. In 1774 Scheele discovered barium, manganese (isolated the same year by Swedish mineralogist Johan Gahn, 1745–1818), and chlorine, although he did not recognize chlorine as an element at first. While studying arsenic in 1775, he prepared copper arsenide, which soon became used as a pigment called Scheele's green. He also noted the effect of light in darkening silver salts, a phenomenon that was to play an important part in the development of photography.

His element score rose to six with the discoveries of molybdenum (1778) and tungsten (1781). A year later he first made the deadly poisonous hydrocyanic acid (prussic acid). Scheele also experimented in organic chemistry. In 1770 he discovered tartaric acid, in 1774 methanoic (formic) acid, in 1776 uric acid, and in 1783 citric acid and glycerol.

Smeaton, John (1724–92)

John Smeaton was an English engineer, best remembered for building lighthouses. He was born near Leeds, the son of a lawyer. He was educated at Leeds Grammar School and in 1740 joined his father's office. At the age of 18 he went to London to continue studying law but instead apprenticed himself to a scientific instrument maker (with his father's consent) and opened his own shop in 1748. His reputation as an engineer grew, and he was elected to the Royal Society in 1753. In 1755 he traveled through Holland to study drainage works and harbors.

In 1756 Smeaton was given the job of building a lighthouse on the Eddystone rocks 14 miles (23 km) off the English coast near Plymouth. (The first Eddystone lighthouse had been swept away in 1703; the second was built of wood and had been destroyed by fire in 1755.) Smeaton used blocks of Portland stone pegged to each other and the foundation, bonded together with hydraulic cement (invented by Smeaton in 1756), which sets under water. Completed in 1759, the lighthouse stood for nearly 120 years before it had to be replaced. The Forth and Clyde Canal in Scotland was another of Smeaton's major engineering works. Begun in 1768, it had 39 locks along its 38-mile (61-km) length. He was also chief engineer on the construction of Ramsgate Harbor in 1774.

In 1767 he began studying model atmospheric steam engines to improve their efficiency—for example, by boring the cylinders more accurately—and he designed a boring machine for this purpose. One of his full-size engines constructed at a coal mine in the north of England in 1772 was 27 percent more efficient (in terms of fuel consumed) than previous engines. However, these advances were soon overtaken by the work of *James Watt* (1736–1819). Smeaton also designed windmills and watermills, building 44 of them between 1753 and 1790.

Stahl, Georg (1660–1734)

Georg Ernst Stahl was a German chemist and physician who was responsible for the phlogiston theory of combustion, an incorrect idea that was to be one of the dead ends of scientific theory. Born in Ansbach in central Germany (Prussia), the son of a Protestant minister, he studied medicine at the University of Jena. He graduated in 1684, became court physician to the duke of Sachsen-Weimar in 1687, and seven years later went to teach at the new University of Halle. From 1716 he was court physician to Frederick-William I, king of Prussia.

Stahl developed the phlogiston theory from the ideas put forward by his compatriot Johann Becher (1635–82) in 1669. He proposed that when a substance burns, it loses a "vital essence," a combustible element called phlogiston. For example, a metal loses phlogiston to form a calx (oxide); but on further heating with charcoal, it recombines with phlogiston to form the metal again. Charcoal must be particularly rich in phlogiston because it leaves hardly any calx on burning. Air, necessary for burning, absorbed the phlogiston as it was evolved. The phlogiston theory dominated chemistry in the latter part of the 18th century until it was finally disproved by the experimental work of **Antoine Lavoisier** (1743–94).

Torricelli, Evangelista (1608–47)

Evangelista Torricelli was an Italian physicist and mathematician who invented the mercury barometer, finally establishing beyond all doubt the existence of atmospheric pressure. He was born in northern Italy and probably became an orphan at an early age, but by 1627 he was receiving education in Rome at the Sapienza College. He drew inspiration from the work of *Galileo* (1564–1642) and in 1641 went to work and live with the great Italian scientist in Florence. When Galileo died the next year, Torricelli took over as professor of mathematics at Florence and stayed there until he died.

Faced with the problem of why the suction pump in the duke of Tuscany's well could not raise the water more than 29.5 feet (9 m), Torricelli concluded that the atmosphere must have weight, but only enough weight to exert a pressure equivalent to a 29.5-foot (9-m) water column. It was the weight of the atmosphere pressing on the surface of water in the well that allowed the pump to lift the water. He chose a much denser liquid—mercury—to test his idea. The experiment, probably performed in about 1645 by his pupil Vincenzo Viviani (1622–1703), involved taking a glass tube, filling it with mercury, and inverting it in a dish of mercury. The level of mercury in the tube dropped to about 30 inches (76 cm) long and remained apparently suspended by an unseen force. Torricelli argued that the "force" was atmospheric pressure pressing down on the surface of the mercury in the dish. The space in the tube above the mercury was a vacuum—the Torricellian vacuum.

Torricelli also noticed that the height of the mercury column tended to vary from day to day according to variations in atmospheric pressure. The mercury column was therefore a measure of this pressure—it was a barometer, known at first as a Torricellian tube. He went on to demonstrate that atmospheric pressure varies with altitude, and that a barometer, suitably calibrated, could be used as an altimeter. He also constructed telescopes and even a simple microscope, and made significant contributions to the mathematics of conic sections. Torricelli's name is commemorated in the unit of pressure called the torr: 1 torr equals 0.04 inches (1 mm) of mercury.

GLOSSARY

Any of the words printed in SMALL CAPITAL LETTERS can be looked up in this Glossary.

Abbreviations
b. Born (of a deceased person for whom only the birth date is known).
c. About (from Latin *circa*).
d. Died (of a deceased person for whom only the death date is known).
date with dash (–), e.g., Bill Gates 1955– , used to signify that, at the time of writing, the person was still living.
fl. Flourished (from Latin *floreat*, of a person whose birth and death dates are not known).

algebra The branch of mathematics that studies the properties of mathematical structures, and in which unknown quantities are denoted by letters.

asteroid Any of thousands of small, rocky bodies that ORBIT THE SUN, mostly between Mars and Jupiter.

astronomy The study of objects outside the Earth's atmosphere.

atom The smallest unit that MATTER can be divided into and still retain its chemical identity.

barometer An instrument for measuring atmospheric (air) pressure.

base number In a system of writing numerals, a number that represents the number of units in a given digit's place that is required to give the numeral 1 in the next higher place. The decimal system that we use today calculates to the base 10, in which 87 is 8 "tens" plus 7 "units."

base 10 *See* BASE NUMBER.

chronometer An extremely accurate mechanical clock formerly used for finding a ship's position (LONGITUDE) at sea.

comet A small icy body in an ORBIT around THE SUN. Some comets, such as Halley's, are short-period comets with highly elliptical orbits bringing them into the inner SOLAR SYSTEM roughly every 100 years. Long-period comets are thought to fall inward from the OORT CLOUD; in some the orbit may be affected by GRAVITATIONAL disturbance from the giant planets, causing them to wander the depths of interstellar space.

compass A device with a pivoted magnetized needle that always swings around to point to MAGNETIC NORTH.

diffraction The fanning out of a wave (e.g., of LIGHT) when it passes through or is reflected from a very fine mesh or grid.

electromagnetism The phenomenon by which magnetic fields can be produced by the flow of ELECTRONS in an electric current.

electron A negatively charged SUBATOMIC PARTICLE.

fossils The remains from prehistoric times of animals or plants preserved in natural materials, such as SEDIMENTARY ROCK, mud, amber, or coal.

galaxy A large collection of STARS, dust, and GAS held together by the force of GRAVITY.

gas A state of MATTER in which the MOLECULES move at random.

geology The group of sciences concerned with the study of the Earth, including its structure, long-term history, composition, and origins.

geometry The branch of mathematics that studies the properties of shapes.

gravity (adj. gravitational) One of the fundamental forces of nature, the force of attraction existing between all MATTER in the UNIVERSE. The gravitational attraction of THE SUN keeps the planets in their ORBITS, and gravitation holds the matter in a STAR together.

latitude The distance of a point on the globe north or south of the equator, measured as an angle. The equator has a latitude of 0°. (*See* LONGITUDE.)

lens A piece of polished glass or transparent plastic shaped in a way that it modifies, or "bends," rays of LIGHT that pass through it. A concave (diverging) lens has a surface that is hollow like the inside of a spoon bowl; a convex (converging) lens bulges outward, like the back of a spoon.

light The form of electromagnetic radiation to which the human eye is sensitive; it is considered to have both particle and wave properties, and the fundamental particle (or quantum) of light is called the photon.

lightning conductor A metal rod placed on a tall structure to protect it by attracting lightning strikes and carrying the charge harmlessly to the ground.

logarithms In mathematics the logarithm of a number is the power to which the base (of the logarithm) must be raised to equal the number, e.g., in common logarithms to the BASE 10, the log of 2 is 0.3010 (because $10^{3010} = 2$).

longitude The distance of a point on the globe east or west of the PRIME MERIDIAN, measured as an angle. (*See* LATITUDE.)

magnetic north The direction indicated by the needle of a COMPASS. It is not exactly the same as true north and changes position slightly over the years.

magnetism The phenomenon associated with magnetic fields, which are produced by moving charged particles. In electromagnets, ELECTRONS flow through a coil of wire connected to a battery; in permanent magnets, spinning electrons within ATOMS generate the field.

mass The amount of MATTER in an object; alternatively, a measure of the extent to which an object resists acceleration when a force is applied to it.

matter Material substance that occupies space and has MASS.

molecule A combination of at least two ATOMS that forms the smallest unit of a chemical element or compound.

Moon, the The Earth's only natural satellite.

moon A natually occurring, relatively large body in ORBIT around a planet.

nebula A cloud of dust and GAS in space that is visible to observers on Earth because it emits, reflects, or absorbs starlight.

neutron One of the three main SUBATOMIC PARTICLES. They carry no electric charge and they occur in the NUCLEI of all ATOMS except hydrogen.

nuclear fusion A process by which two or more ATOMIC NUCLEI join together to make a heavier one.

nucleus, atomic (pl. nuclei) The positively charged dense region at the center of an ATOM, composed of PROTONS and NEUTRONS.

Oort cloud A spherical region, believed to surround the SOLAR SYSTEM, containing a vast number of COMETS.

orbit The path that an object takes under the influence of GRAVITY.

prime meridian The zero line of LONGITUDE (0°) that passes through Greenwich, England, and from which longitude and time are measured.

prism A usually triangular block of a transparent material that can split white LIGHT into the colors of the rainbow.

proton A positively charged SUBATOMIC PARTICLE in the ATOMIC NUCLEUS.

reflecting telescope A TELESCOPE that forms an image using a mirror (not a LENS).

refracting telescope A telescope that forms an image using a LENS (not a mirror).

retina A light-sensitive membrane lining the back of the eye that senses light and transmits impulses along the optic nerve to the brain.

sedimentary rock Rock formed from organic or mineral fragments and deposited in layers by ice, water, or wind.

solar system Everything that is dominated by THE SUN's GRAVITATIONAL field. It is made up of the Sun, the nine planets and their MOONS as well as minor bodies such as ASTEROIDS and COMETS.

star A celestial object that shines because of the release of energy liberated in its core by NUCLEAR FUSION.

star, binary A stellar system composed of two STARS that ORBIT one another about their common center of MASS. They are held together by the force of their mutual GRAVITY.

star, variable Any STAR whose luminosity (brightness) varies.

subatomic particle Any particle that is smaller than an ATOM.

Sun, the The STAR at the center of our SOLAR SYSTEM, around which the Earth and other planets ORBIT.

sun a celestial body that resembles the SUN. (*See* SUN, THE.)

telescope An optical instrument that produces an enlarged image of distant objects. (*See* REFLECTING TELESCOPE; REFRACTING TELESCOPE.)

Universe The whole of space, time, and everything in it.

vacuum A completely empty space in which there are no atoms or molecules of any substance.

SET INDEX

FURTHER READING

General

Allaby, M., and D. Gjersten, *Makers of Science*, Oxford University Press, 2002.

Asimov, A., *Asimov's Biographical Encyclopedia of Science and Technology*, Avon Books, 1972.

Boorstin, D. J., *The Discoverers*, Random House, 1983.

Boyles, D., *The Tyranny of Numbers*, HarperCollins, 2000.

Bunch, B., and A. Hellemans, *The Timetables of Science*, Simon and Schuster, 1988.

Bunch, B., and A. Hellemans, *The Timetables of Technology*, Simon and Schuster, 1993.

Carey, J. (ed.), *The Faber Book of Science*, Faber and Faber, 1995.

Crystal, D. (ed.), *The Cambridge Biographical Dictionary*, Cambridge University Press, 2000.

Daintith, J. (ed.), *A Dictionary of Scientists*, Oxford University Press, 1999.

Day, L., and I. McNeil (eds.), *Biographical Dictionary of the History of Technology*, Routledge, 1998.

Diamond, J., *Guns, Germs and Steel,* Vintage, 1998.

The Dictionary of National Biography, Oxford University Press, 1982.

Dyson, J., and R. Uhlig (ed.), *A History of Great Inventions*, Robinson, 2002.

le Fanu, J., *The Rise and Fall of Modern Medicine*, Little, Brown and Company, 1999.

Giscard d'Estaing, V-A., *The Book of Inventions and Discoveries*, Macdonald Queen Anne Press, 1990.

Gribbin, J., *Science, A History*, BCA, 2002.

Harrison, I., *The Book of Inventions*, Cassell, 2004.

Hoskin, M. (ed.), *The Cambridge Concise History of Astronomy*, Cambridge University Press, 1999.

The Hutchinson Dictionary of Scientific Biography, Helicon, 2000.

The Inventions that Changed the World, Reader's Digest, 1982.

McLeish, J., *Number*, Bloomsbury, 1991.

Margotta, R., *The History of Medicine*, Octopus, 1996.

Meadows, J., *The Great Scientists*, Oxford University Press, 1997.

Messadié, G., *Great Inventions through History*, Chambers, 1991.

Messadié, G., *Great Scientific Discoveries*, Chambers, 1991.

Millar, D., et al., *The Cambridge Dictionary of Scientists*, Cambridge University Press, 1996.

Muir, H. (ed.), *Larousse Dictionary of Scientists*, Larousse, 1994.

Parry, M. (ed.), *Chambers Biographical Dictionary*, Chambers, 1997.

Philip's Astronomy Encyclopedia, George Philip Limited, 2002.

Philip's Science & Technology Encyclopedia, George Philip Limited, 1998.

Philip's Science & Technology People, Dates & Events, George Philip Limited, 1999.

Porter, R., *The Greatest Benefit to Mankind*, HarperCollins, 1997.

Silver, B. L., *The Ascent of Science*, Oxford University Press, 1998.

Tallack, P (ed.), *The Science Book*, Cassell, 2001.

Trefil, J., *Cassell's Laws of Nature*, Cassell, 2002.

Waller, J., *Fabulous Science*, Oxford University Press, 2002.

Webster's Biographical Dictionary, G. & C. Merriam, 1971.

What Happened When?, HarperCollins, 1994.

Whitfield, P., *Landmarks in Western Science*, The British Library, 1999.

Williams, T. I. (ed.), *A Biographical Dictionary of Scientists*, Adam & Charles Black, 1974.

Williams, T. I., *A History of Invention*, Time Warner Books, 2003.

Specific to this Volume

Brockman, J. (ed.), *The Greatest Inventions of the Past 2,000 Years*, Weidenfeld & Nicolson, 2000.

Clark, D. H., and S. P. H. Clark, *Newton's Tyranny*, W. H. Freeman and Company, 2001.

Emsley, J., *The Shocking History of Phosphorus*, Macmillan, 2000.

Inwood, S., *The Man Who Knew Too Much*, Macmillan, 2002.

Lane, N., *Oxygen: The Molecule that Made the World*, Oxford University Press, 2003.

Martin, S., *Alchemy and Alchemists*, Pocket Essentials, 2001.

Peterson, H. L., *The Book of the Gun*, Paul Hamlyn, 1962.

Sobel, D., *Galileo's Daughter*, Fourth Estate, 1999.

Thompson, L., *Guns in Colour*, Octopus, 1981.

Woodman, R., *The History of the Ship*, Conway Maritime Press, 1997.

Terminology Reference

Clark, J. O. E., and S. Stiegler (eds.), *The Facts on File Dictionary of Earth Science*, Checkmark Books, 2000.

Daintith, J. (ed.), *A Dictionary of Chemistry*, Oxford University Press, 2000.

Daintith J., and J. Clark (eds.), *The Facts on File Dictionary of Mathematics*, Facts on File, 1999.

Darton, M., and J. Clark, *The Macmillan Dictionary of Measurement*, Macmillan, 1994.

Illingworth, V., *Collins Dictionary of Astronomy*, (ed.), HarperCollins, 1994.

Waites, G., *The Cassell Dictionary of Biology*, Cassell, 1999.

USEFUL WEB SITES

http://www.atomicmuseum.com/
Web site of the National Atomic Museum (soon to be renamed the National Museum of Nuclear Science and History), Albuquerque, New Mexico.

http://www.howstuffworks.com
A comprehensive Web site that gives detailed explanations of how everything around us actually works.

http://www.mhs.ox.ac.uk/
Web site of Oxford University's Museum of the History of Science.

http://www.nasa.gov/
Official Web site of the National Aeronautics and Space Administration (NASA) with links to learning resources and information for students, and full details of all missions.

http://www.nasm.si.edu/
Web site of the Smithsonian National Air and Space Museum, Washington, D.C.

http://www.nscdiscovery.org/
Web site of the National Science Center, Augusta, Georgia.

http://nobelprize.org/
Official Web site of the internatonal Nobel Prize awards given annually since 1901 for achievements in physics, chemistry, medicine, literature, and for peace. Click on "list of all prizewinners" and then on individual links to read about award-winning achievements.

http://www.psigate.ac.uk/
Physical Sciences Information Gateway provides access to Web resources in the physical sciences, including astronomy, chemistry, earth sciences, materials sciences, physics, and general science.

http://whyfiles.org/
University of Wisconsin Web site that explains the science behind the news.